This introductory account of our Universe is based on the author's Christmas Lectures, given at the Royal Institution over the New Year holiday 1990–91. The Royal Institution lectures for young people are televised to a live audience. This book brilliantly captures Professor Longair's flair as a skilled presenter of modern astronomy and physics. The contents include chapters on The Grand Design, The Birth of Stars and the Great Cosmic Cycle, The Origin of Quasars, The Origin of Galaxies, and The Origin of the Universe. The language is non-technical and the book is well-illustrated.

The Origins of our Universe

The Origins of our Universe

A Study of the Origin and Evolution
of the Contents of our Universe

The Royal Institution Christmas
Lectures for Young People

1990

Malcolm S. Longair

Published by the Press Syndicate of the University of Cambridge
The Pitt Building, Trumpington Street, Cambridge CB2 1RP
40 West 20th Street, New York, NY 10011–4211, USA
10 Stamford Road, Oakleigh, Victoria 3166, Australia

First published 1991
Reprinted 1992

Printed in Great Britain at the University Press, Cambridge

British Library cataloguing in publication data available

Library of Congress cataloguing in publication data available

ISBN 0 521 41140 8 hardback
ISBN 0 521 42303 1 paperback

For Mark and Sarah

Contents

Preface

It is a great privilege to have been invited to present the 1990 Royal Institution Christmas lectures for Young People. The remarkable tradition established by Faraday in the 19th century and continued into the present century under the direction of scientists of the highest distinction demands a special effort. It is a unique opportunity to bring to the next generation of scientists and engineers the excitement of contemporary science. My theme is the origins of everything there is in the Universe, one of the greatest challenges of modern science. I cannot hope to more than whet the appetite of the enthusiast but what I do aim to get across is the intellectual excitement and the extraordinary demands of research at the frontiers of our present understanding.

To aid the dissemination of these ideas, I have produced this short book which is intended to be no more than extended lecture notes for the series of five lectures. I hope they will fill in some of the gaps in the televised lectures and act as a stimulus for those interested to read and learn more.

This book could not have been produced without very special efforts on the part of many people. First and foremost is Professor John Thomas, Director of the Royal Institution, who invited me to give this series of lectures. Sponsorship for the series is provided by Shell whose generosity and encouragement I acknowledge. Dr Bryson Gore of the RI has provided invaluable advice on the lectures and discouraged me from excessively original experiments. Mr William Woollard of InCA and my director Mr Richard Melman have provided wonderfully creative ideas concerning the presentation of this material. The support of Graham Massey and his staff at the BBC, who broadcast the lectures, is also much appreciated.

Cambridge University Press has pulled out all the stops to make this book available by the time of the Christmas lectures. I am very grateful to Dr Simon Mitton of CUP for his enthusiasm and to his staff for doing their job at a superluminal speed. Finally, I could not have made the deadlines without the terrific efforts of Mrs Janice Murray of the Royal Observatory, Edinburgh, in helping produce the book in TEX. Her efforts even exceeded her normal enormous output of work.

Malcolm S. Longair
Royal Observatory
Edinburgh

1

The Grand Design

In this short volume we are going to study the origin and evolution of everything there is in the Universe. We will use all the techniques of modern astronomy to ask some of the deepest questions about the origins of planets, stars, galaxies, larger scale structures and the Universe itself. We have a great deal to do. In this first chapter, we have to understand not only what it is we are trying to explain but also the ways in which we can obtain the information we need. So, let's begin with a description of what our Universe looks like.

The first thing I am very often asked by non-astronomers is, "How can you ever imagine the huge sizes and distances involved in the study of the Universe? How can the human mind comprehend such vast distances?" The answer is that it really is very easy because you must not worry about the size of the numbers themselves - that is not really important at all. What is important is the size of one scale in the Universe relative to another. So, let's begin with a quick run through the Universe - a Grand Tour, if you like - starting from our local Universe and getting as quickly as possible to the edge of the Universe. Of course, I have to tell you what I mean by the edge of the Universe but let's wait until we get there.

We will start right in our back garden. I call our back garden those regions of the Universe to which we have already sent space vehicles. Much of our own Solar System has already been explored by space probes such as the Voyager I and II spacecraft which provided such

Figure 1. Saturn as observed by the Hubble Space Telescope.

fantastic views of the outer planets - Jupiter, Saturn, Uranus, Neptune - and their satellites.

You are all familiar with the idea that the planets revolve in more or less circular orbits about the Sun. What is important for our present purposes is the size of the Solar System relative to the size of larger systems in the Universe and so let's begin that journey. First of all, we compare the distance from the Sun to the Earth with the distance from the Sun to Saturn (Figure 1). That is only a factor of 10 in distance and 10 is a small number. Although the actual distance in kilometres from the Sun to Saturn might seem to be very large, when you think of it in relative terms as only 10 times the distance from the Sun to the Earth, it is not so enormous. This is the first step in a sequence which will take us out to the edge of the Universe.

The next stage in our journey is to travel from our Sun with its nine planets to the nearest neighbouring star. How far do we have to go? If we were to write down the distance in kilometres it would seem a very large number but, if we think of it in terms of the distance of the Sun to Saturn, it is only about 20 000 times that distance to get to the nearest star. Again, 20 000 is not a particularly large number. Another way of thinking about the distance of the nearest stars is to work out the time

Figure 2. The Pleiades star cluster. These blue stars are members of a young star cluster, no more than 20 million years old, and lie at a distance of about 300 light years.

it takes light to get there from the Sun. Light travels at a speed of about 300 000 kilometres per second. Thus, it takes light only about 8 minutes to travel from the Sun to the Earth. Correspondingly, it takes light about 4.3 years to travel from the Sun to the nearest star, Alpha Centauri - we say that Alpha Centauri is at a distance of 4.3 light years. To give you some idea of how many stars there are nearby, we know of about 50 stars within a distance of 17 light years of the Sun.

In our neighbourhood, most of the stars are cool red stars with a few compact hot blue stars known as white dwarfs. Our own Sun is a very average star. The masses of the nearby stars all lie within a quite narrow range, from about two times to about one-tenth the mass of the Sun. To find the most luminous stars such as those in the Pleiades (Figure 2), we have to search much larger volumes of the Galaxy because these are rather rare stars, having very short lives compared with the nearby stars.

Now let's go up another scale in size. Suppose we go up another 30 000 times the distance between the stars - then we obtain a dramatically different perspective. The stars are congregated into enormous "island universes" - what we call the **galaxies**. Our own Galaxy, the Milky

Figure 3. The nearest neighbouring giant spiral galaxy to our own Galaxy, the Andromeda Nebula or M31.

Way, is probably very like the Andromeda Nebula, our nearest spiral neighbour in space (Figure 3). The nearby stars are only a few of the billions of stars which make up the disc of our Galaxy. The stars of the disc rotate about the centre of the Galaxy and it is the balance between the gravitational attraction of the Galaxy as a whole and the centrifugal force due to its rotation which holds the stars of the disc in more or less circular orbits. In addition, there is a prominent central bulge of stars. Altogether our Galaxy must contain about 100 000 000 000 (one hundred billion or 10^{11}) stars, a very, very large number indeed.

The problem with observing our own Galaxy is that we are located inside it and so we cannot obtain a bird's eye view of it as we can for distant galaxies. In fact, the Milky Way, which you can only see on a very clear dark night from Great Britain, is simply the disc of our own Galaxy seen edge-on. It does not look very like the Andromeda Nebula, however, because, as can be seen in Figure 4, many regions are obscured

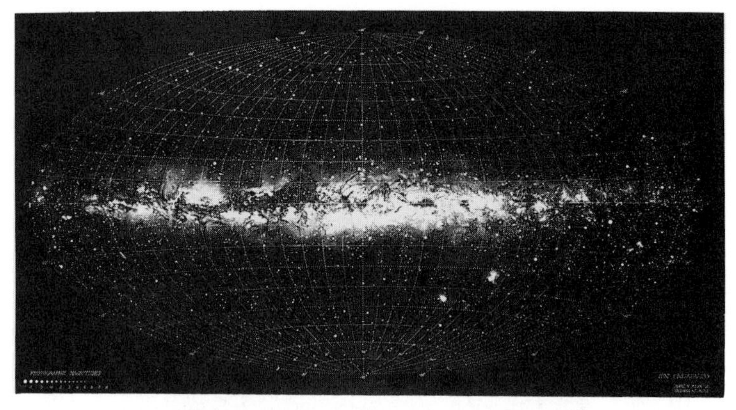

Figure 4. A view of the Milky Way showing the whole of the sky projected onto a two-dimensional map. In this image, the Milky Way runs across the middle of the diagram with the centre of the Galaxy in the centre of the picture. The coordinates have been squashed so that equal areas are preserved. This picture shows that our Galaxy is a flattened disc but the presence of dust prevents us having a clear view of its structure.

Figure 5. An image of our own Galaxy as observed by the Cosmic Background Explorer (COBE) in the infrared wavebands, 1 to 2.2 μm.

by interstellar dust. If, however, we observe at infrared wavelengths at which the dust in the disc of the Galaxy becomes transparent, we see very clearly the disc and bulge structure present in galaxies such as the Andromeda Nebula (Figure 5).

Galaxies such as the Andromeda Nebula and our own Galaxy are the

Figure 6. The barred spiral galaxy NGC1365.

building blocks of our Universe. Let us look at some other examples of the population of galaxies.

The vast majority of galaxies are called **normal galaxies** and they come in two broad classes. One of these classes consists of the **spiral galaxies**, like our own Galaxy and M31 as well as those known as **barred spiral galaxies** in which the central bulge is replaced by a ellipsoidal "bar" of stars (Figure 6). The characteristic feature of this class is the presence of beautiful spiral arms. The second major class consists of the **elliptical galaxies** which are spheroidal systems which do not contain discs of stars (Figure 7). There are galaxies which seem to be intermediate between the two classes and these are known as the **S0** or **lenticular galaxies**. Finally, there is a small class of peculiar galaxies known as **irregular galaxies**. Every now and then we find peculiar things happening in the Universe and Figure 8 is an example in which a galaxy seems to have been involved in a crash with one of its neighbours and has left behind a ring of young stars. All the galaxies in the Universe fall into these general categories.

There are some galaxies, similar in size to our own Galaxy and greater, in which much more dramatic events are taking place. A good example

Figure 7. The elliptical galaxies NGC1399 and 1404 which lie in the Pavo cluster of galaxies.

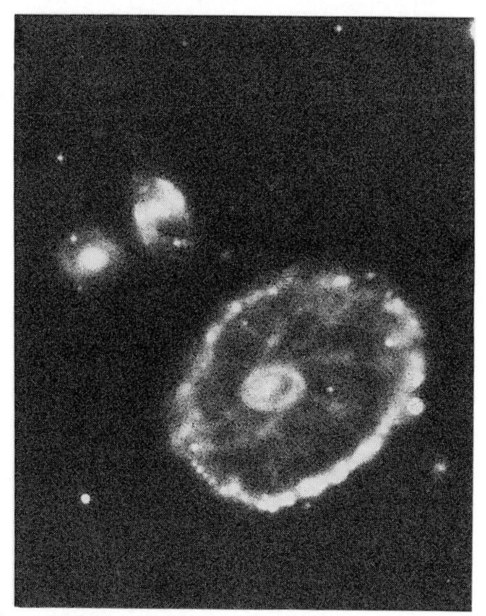

Figure 8. The peculiar galaxy known as the Cartwheel. Its strange appearance is almost certainly due to a recent collision or to an interaction with one of its nearby companions.

of this is the galaxy NGC4151. At first sight, this galaxy appears to be rather harmless (Figure 9). When we take a short exposure photograph, however, something very dramatic seems to be happening in its nucleus (Figure 10). There seems to be a star located in its nucleus but it cannot be any normal sort of star at all because it is at the same distance as the galaxy. What makes matters even more intriguing is the fact that this star-like object is known to be variable in intensity which, as we will show in Chapter 3, means that it must be very compact. This is just one example of a class of galaxies which possess **active galactic nuclei**. In the example of NGC4151, the active nucleus is much fainter than the galaxy but in very, very rare cases, the opposite is found - the nucleus far outshines the galaxy and these are the objects known as **quasi-stellar objects** or **quasars**. Figure 11 is a photograph of the brightest quasar in the sky, 3C 273. The faint smudges towards the bottom of the picture are companion galaxies at the same distance as the quasar. Evidently, the nucleus outshines the underlying galaxy by a factor of about 1000 in luminosity. Almost certainly the quasars are the most powerful sources of energy that we know of anywhere in the Universe. We will have a great deal to say about them later but for the moment we note that most galaxies are more or less normal and that only very rarely do we find really exotic objects which have quite different properties.

Let us go up again in scale. We have already stated that the galaxies are the building blocks of the Universe. To make a map of the Universe on the very largest scale, we plot out where all the galaxies are located in the Universe. It turns out that the galaxies are far from randomly distributed in the Universe. They generally form rather irregular structures as we will see but there are some large systems of galaxies, the largest bound systems that we know of in the Universe, which are known as **clusters of galaxies**. In the great clusters of galaxies, there may be many thousands of galaxies and the cluster is held together by the mutual gravitational attraction of the galaxies for one another. Figure 12 shows the Pavo Cluster of galaxies and our own Galaxy would not be a particular conspicuous member, certainly nothing like the giant monster galaxy in the centre of the cluster. The size of the whole cluster is about 50 to 100 times the size of our Galaxy. We are now getting up to quite large scales.

Now what about even larger scales? This is one of the most exciting and important areas of modern cosmology. The distribution of galaxies on larger scales does not seem to be particularly smooth. To appreciate what these structures look like, Figure 13 is a picture of the whole of the Northern Galactic Hemisphere produced by James Peebles and his

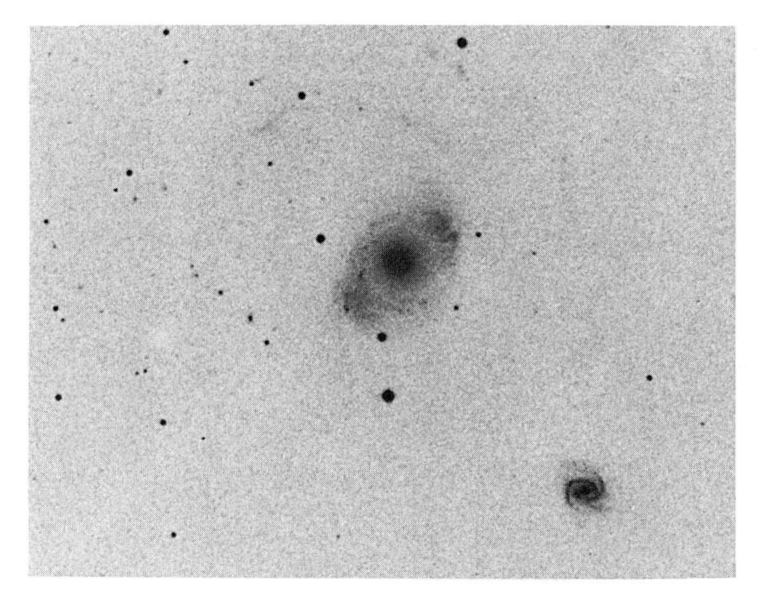

Figure 9. A long exposure photograph of the galaxy NGC4151. The galaxy appears to be a spiral galaxy.

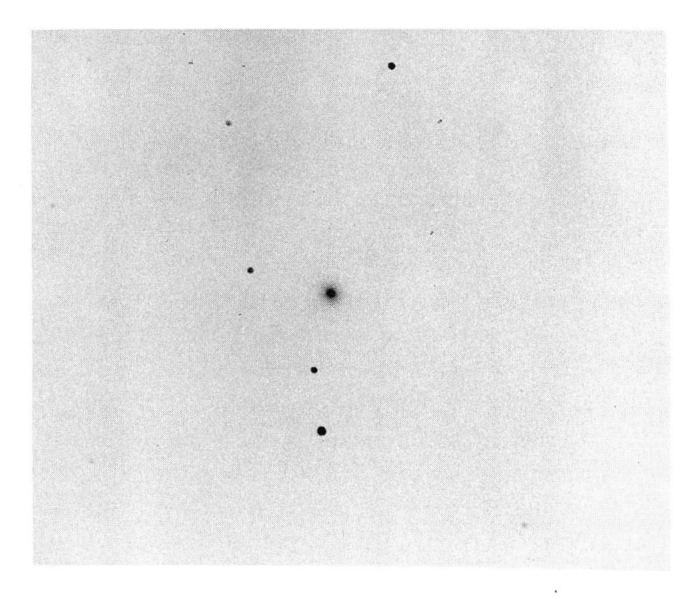

Figure 10. A short exposure photograph of NGC4151. The nucleus of the galaxy appears to be stellar. The fact that the nucleus is very compact is confirmed by the observation that its brightness varies.

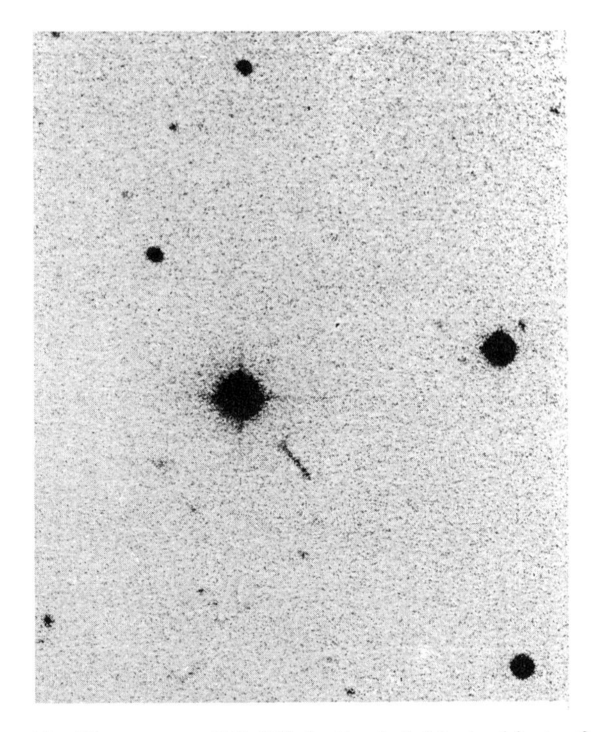

Figure 11. The quasar 3C 273 is the brightest object of this class known. Although it looks like a star, it is in fact very distant indeed. The faint smudges seen beneath the quasar are galaxies at the same distance as the quasar. The nucleus outshines the galaxy by a factor of about 1000 in luminosity.

colleagues at Princeton University with all the stars in our own Galaxy removed - in other words, it shows where the galaxies are located on the sky. This is a view of the Universe on the very grandest scale. Individual clusters are simply the peaks in the distribution of galaxies. You can see that the galaxies do not seem to be uniformly distributed - there are holes, filaments and sheets of galaxies. This large scale structure is most beautifully illustrated by the results of the Harvard-Smithsonian Astrophysical Observatory Survey of Galaxies. In this survey, distances as well as positions on the sky have been measured for over 20 000 galaxies. Figure 14 shows the three-dimensional picture of the Universe produced by Margaret Geller, John Huchra and their colleagues from part of that survey. You will notice that we can see large clusters of galaxies as well as sheets and filaments of galaxies and great holes in the distribution of

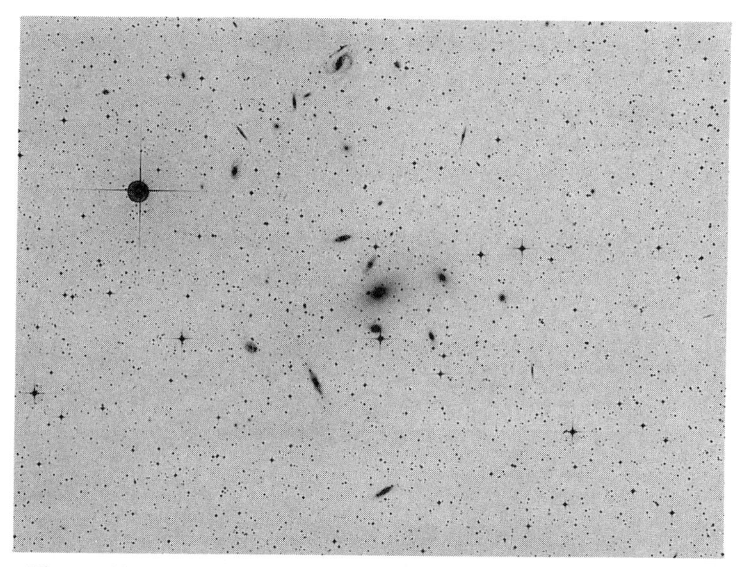

Figure 12. The Pavo cluster of galaxies. In this image, our own Galaxy would be similar to the spiral galaxies seen towards the bottom of the photograph.

galaxies which are often called voids. It is a really remarkable illustration of gross irregularities in the distribution of galaxies in the Universe.

What is the best way of describing this distribution of galaxies? Richard Gott and his colleagues at Princeton have provided us with an elegant and simple answer. You can think of the Universe as being rather like a sponge. You are well aware of the fact that the material of the sponge is all joined together or else it would fall apart. What is much more remarkable is the fact that the holes are all joined together too! This is what happens in the Universe too. The holes are all connected and the galaxies (material of the sponge) are all connected as well. The rich clusters of galaxies tend to lie in the vertices of the sponge, regions where various bits of the sponge come together. Now how big are these holes or voids? It turns out that the largest that have been detected so far are about 100 times the size of a cluster of galaxies. The origins of the holes, sheets and strings of galaxies are among the greatest challenges facing the theorists and we will explain in Chapter 4 why they pose such a nasty problem for theory. For the moment, however, we are only interested in describing the Universe.

Now, what about the size of Universe itself. Put very simply, the size of the Universe is only about 50 times the size of one of the large holes

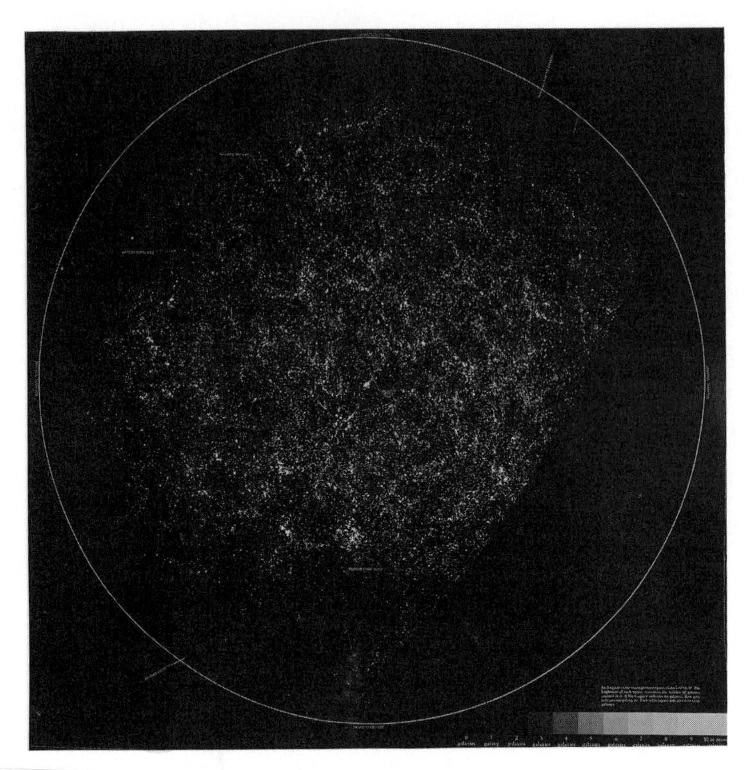

Figure 13. A picture of the distribution of galaxies in the Northern Galactic Hemisphere. The centre of the diagram corresponds to looking vertically out of the plane of our Galaxy, and the outer white circle corresponds to looking through the Galactic Plane. The coordinate system has been chosen so that equal areas are preserved. The lack of galaxies towards the edge of the diagram is due to obscuring dust in the plane of our Galaxy. The "bite" out of the diagram at the bottom right occurs because that area of sky was not surveyed.

or voids. This may surprise you. The reason that this is not truly a surprise is that we have to be careful about what we mean by the size of the Universe. We now have to introduce one of the most useful concepts for understanding how we can study the origin and evolution of the Universe. We are going to have a great deal to say about light in just a moment. Light travels at a very great speed and it always has the same high value when it is propagated through a vacuum. This means that as we look further and further away, it takes longer and longer for the light to reach to us. So, when we look at very distant objects, we are not

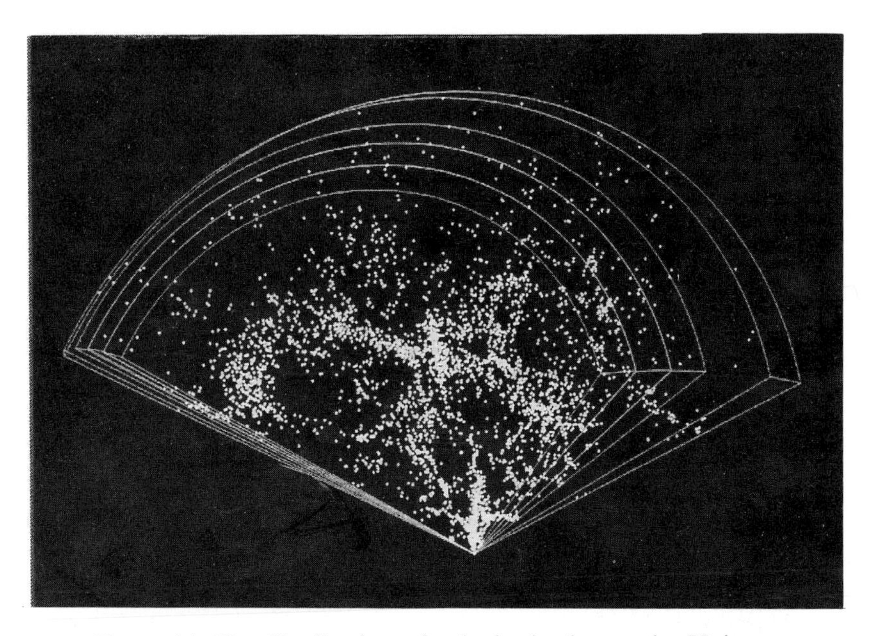

Figure 14. The distribution of galaxies in the nearby Universe as derived from the Harvard-CfA survey of galaxies. This picture is a projection of the three-dimensional distribution of galaxies. Our Galaxy is located at the apex of the segments in which the distances of the galaxies have been measured. The distribution is grossly non-uniform with huge sheets, filaments and voids in the distribution of galaxies.

really seeing the Universe as it is now but as it was when the light was emitted from these objects, which was in the distant past. What I mean by the size of the Universe is the distance to which we can observe the Universe more or less as it is now.

Let me try to make this a little bit more concrete. We will show in Chapters 4 and 5 that we live in a Universe which has expanded from a very compact state indeed and that all the galaxies are moving away from that initial explosion which we call the Big Bang. Now, if we go, not back to the beginning, but only to the time when the Universe was half its present age, we are no longer looking at the Universe as it is now because it was much younger then and will certainly be significantly different from the Universe now. Just think what you were like when you were only half your present age and you will realise that objects evolve considerably over a factor of two in time. So what we mean by the size of the Universe now, is the distance out to which we can see

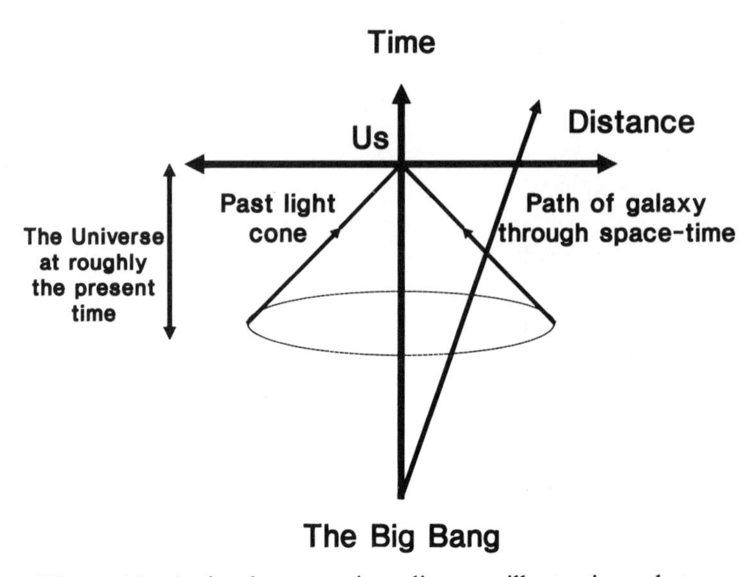

Figure 15. A simple space-time diagram illustrating what we mean by the "the size of the Universe". Time runs up the diagram and distance across it. When we observe distant objects, we look into the past. Eventually, when we look very far away, we see the Universe as it was when it was much younger than it is now.

the Universe when it is more or less its present age and that distance is roughly 50 times the size of the big holes (Figure 15). Now there is lots more Universe beyond that distance but we cannot see it - light cannot get to us from these regions in the time since the light was emitted. If we look further and further away we are indeed looking at more and more distant bits of the Universe but not as they are now but as they were in the very, very distant past. We will study these very early phases later.

We have now travelled from our back yard, our own Solar System, right to the edge of the Universe in only six remarkably modest steps and the remarkable thing is that we have not been talking about huge numbers at all. We have been talking about little numbers which were never any larger than 30 000. This is an interesting scientific point in that when scientists are faced with the problem of having to deal with a very wide range of scales, they do not think linearly in terms of all the zeros on the end of the numbers but rather they think logarithmically, i.e. one scale relative to another. When we think in this way, the Universe is really a rather cosy place.

So far I have shown you a very beautiful but biased picture of our Universe. All the pictures have been taken with optical telescopes and that introduces a strong bias because light waves are only one brand of the waves we can use to look at the Universe. The important concept is the idea that light is only one form of what we call **electromagnetic radiation**, in other words radiation due to oscillating electric and magnetic fields, a discovery made by that great Scottish scientist James Clerk Maxwell who was one of the first Professors of Natural Philosophy at the Royal Institution. Because of the transparency of the atmosphere, light in the optical waveband is a remarkably simple and effective way of studying the Universe. However, the Universe does not particularly care about us or our eyes and in fact astronomical objects generally radiate electromagnetic waves in a very wide range of wavebands in addition to the optical waveband.

We have to understand two important points about these waves. The first thing is that waves carry energy. If you have ever been knocked over by a breaking wave on the seashore, you will know jolly well that waves can push you over. Waves have a certain frequency which I will call f cycles per second, or hertz, as we call them after the German scientist who fully confirmed Maxwell's electromagnetic theory. The waves also have a wavelength which is just the distance between wave crests and I will call this wavelength by the Greek letter lambda, λ. When light or electromagnetic waves, in general, travel through a vacuum, the wavelength and frequency are related to the velocity of light by the simple formula

$$\lambda f = c$$

where c is the velocity of light. Thus, the shorter the wavelength, the greater the frequency of the waves.

The second point is that the shorter the wavelength (or the higher the frequency) of the wave the more energy is transported. If we run through the optical spectrum from red through to blue and ultraviolet wavelengths, we go from lower to higher frequencies of oscillation of the electromagnetic waves or, equivalently, from longer to shorter wavelengths. However, the electromagnetic spectrum extends far, far beyond the red end and far, far beyond the blue end of the optical spectrum.

The lowest frequency waves which are used for astronomy are the **radio waves** and these are the waves that you pick up with a radio receiver or with a television aerial. Nowadays, we make maps of the sky with radio telescopes just as we can with optical telescopes. If we move to somewhat shorter wavelengths, or higher frequencies, we come to the **infrared** waveband. The infrared cameras which help rescue workers to detect warm bodies in catastrophes such as the earthquake in Armenia are simple examples of a device which enable us to "see" in the dark. They also enable gun sights to be developed for aiming at warm targets in the dark. When people talk about night vision, they mean looking at the infrared radiation emitted by warm bodies using infrared cameras. Going beyond the optical waveband, we come to the **ultraviolet** and **X-ray** wavebands. I expect everyone has had an X-ray at one time or another and these are just very high frequency electromagnetic waves which can be used to examine whether or not you have broken bones - this works because the bones absorb X-rays more than skin and muscle. Finally, at the very highest energies there are the **gamma-** or **γ-rays**. Very high energy γ-rays can be detected by Geiger counters and are produced in natural radioactivity and also in nuclear explosions. This catalogue makes the point that electromagnetic waves of a very wide range of frequencies and wavelengths play an important role in everyday life.

The next key concept is the idea that if objects are at a particular temperature, they emit radiation most strongly at a particular wavelength. This was a wonderful and important discovery which was made at the end of the last century - there is a very simple relationship between the frequency at which most of the radiation of a hot body is emitted and its temperature. In fact they are proportional to one another - this is known as **Wien's Displacement Law**, to give it its proper name. It will be useful to write this formula down because it contains so much that is important for astronomy. If T is the temperature of a hot body, then most of the radiation is emitted at a wavelength $\lambda_{\max}$ and frequency $f_{\max}$ which are given by

$$f_{\max} = 10^{11} T \quad \text{Hz} \qquad \lambda_{\max} = \frac{3000}{T} \quad \mu\text{m}$$

where the temperature T is measured in kelvins (K), i.e. temperatures measured relative to the absolute zero of temperature which is about -273 °C. What this means is that, if we see radiation from any object then we know how hot that object has to be in order to be able to radiate radiation at that wavelength. These relations are displayed in Figure 16 where the diagonal line stretching from the bottom left to top right represents Wien's Displacement Law. You can read off this diagram the

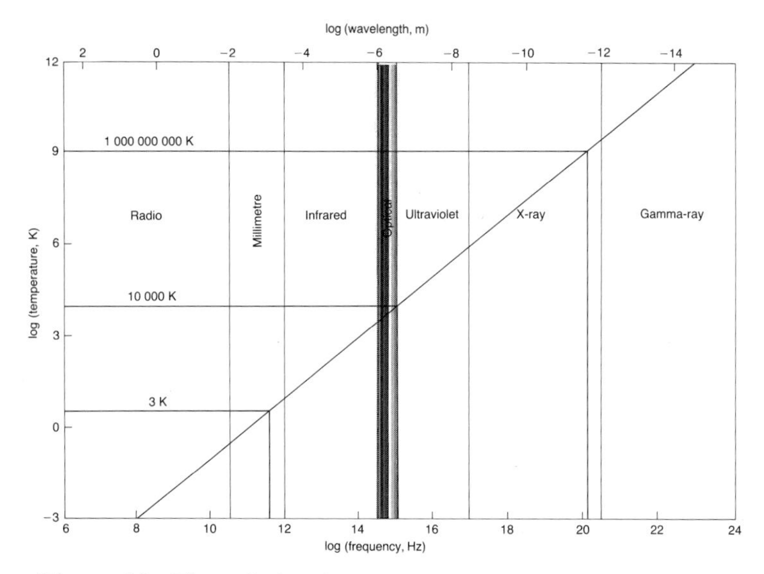

Figure 16. The relation between the temperature of a black body and the frequency (or wavelength) at which most of the energy is emitted. The various wavebands described in the text are labelled.

typical temperature associated with a particular wavelength or frequency. The frequency ranges of the different wavebands are also shown.

The problem for astronomy is that only certain of these wavebands are accessible from ground-based observatories. Figure 17 shows how high a telescope has to be placed above sea level in order to be able to see the sky outside the Earth's atmosphere. You can see that optical and radio astronomy can be carried out from the ground but ultraviolet, X-ray and γ-ray astronomy as well as far infrared astronomy have to be carried out from above the Earth's atmosphere, preferably from satellite observatories.

It turns out that the radiation spectrum of a hot body at a particular temperature has a very simple form, first derived by the famous German scientist Max Planck. This characteristic spectrum is known as black-body radiation and is the spectrum expected of a hot body which is in thermal equilibrium with its surroundings. This means that the body and its surroundings have been left for a very long time so that every process of absorption is balanced by the corresponding emission process. A simple illustration of these results for optical observations is shown in Figure 18. I show the energy distributions radiated by different hot

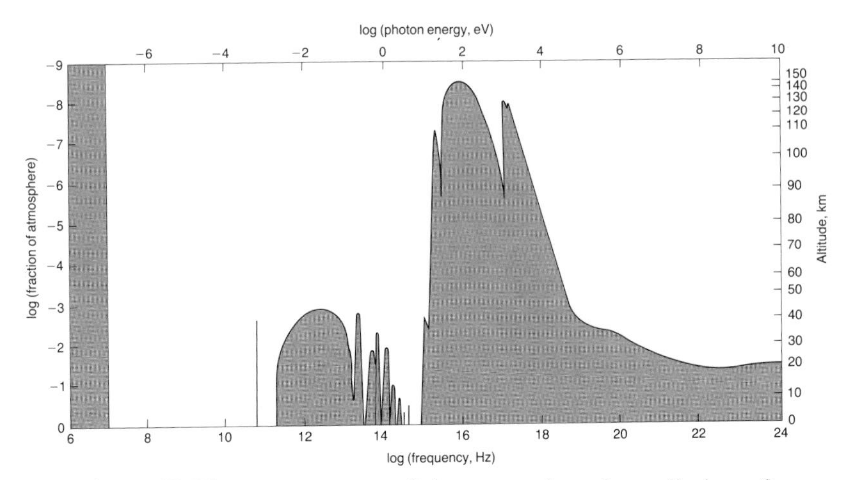

Figure 17. The transparency of the atmosphere for radiation of different wavelengths. The solid line shows the height above sea level at which the atmosphere becomes transparent to radiation of different wavelengths.

bodies in the optical waveband. Now, lots of the radiation is emitted outside this narrow range of frequencies and that is shown as dotted lines at wavelengths outside the optical window. I show the expected thermal distributions of radiation for bodies at temperatures of 3000, 7000, 12 000 and 20 000 K. It can be seen that, as the temperature increases, the maximum of the radiation spectrum moves to shorter and shorter wavelengths. I have highlighted the red, green and blue wavebands of the optical spectrum because these are the familiar colours in it. It can be seen that cool stars are predominantly red, hot stars are predominantly blue and, in between, a star like the Sun which has a temperature about 7000 K, has a roughly equal mixture of red, green and blue colours which makes it more or less white with a tinge of yellow!

This is a very important result because it tells us that when we look at the Universe in different wavebands, either the radio, infrared, optical, ultraviolet, X-ray or γ-ray wavebands, we are looking at different temperature pictures of the Universe. It is as if you took a picture of the Earth and, first of all, took a picture of where all the cold material is (say 0 °C). Then, you take a picture at a temperature of 15 °C and finally you take a picture of the Earth at 25 °C - you can understand

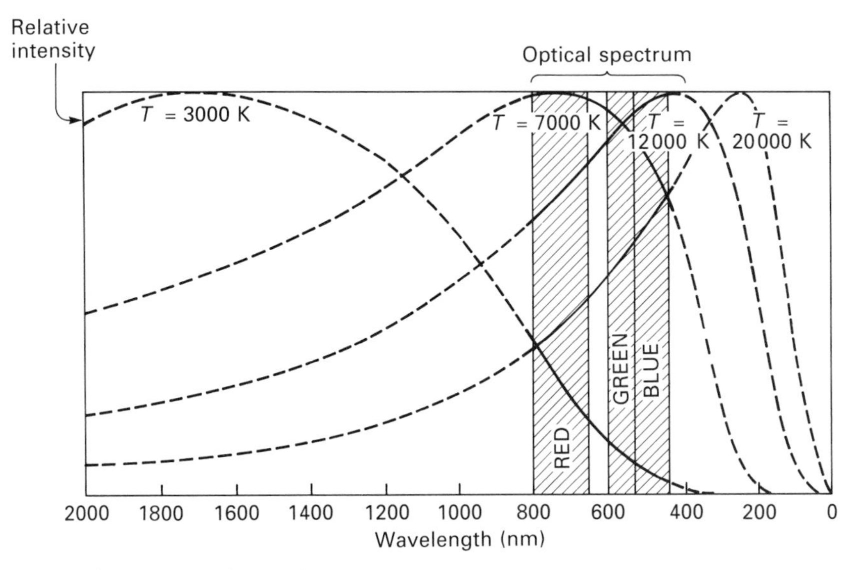

Figure 18. Illustrating the radiation spectra expected of hot bodies at different temperatures. The lines are black-body spectra and the wavelength range shown includes the optical waveband. These spectra have been scaled to have the same maximum intensity.

that we would obtain very different pictures. In modern astronomy, we can look at objects with temperatures which are within a few degrees of absolute zero right up to those with temperatures as high as a billion K and more.

Let us demonstrate how this works for a particular object. The Royal Institution is famous for its explosions and so let us look at an exploding star as seen in different wavebands. Let's look at the remnant of a star which exploded about 250 years ago in the constellation of Cassiopaeia. First of all, we look at it in the standard optical waveband (Figure 19). These straggly optical filaments are seen expanding away from the site of the explosion. These filaments cool by radiating away their energy in the optical waveband because their temperatures lie in the right range. But the object is much more spectacular in the X-ray waveband. Figure 20 shows exactly the same object but now as observed by the Einstein X-ray satellite about 10 years ago. You can see that the remnant contains a sphere of very hot gas. Its typical temperature is about 10 million K since it is observed in the X-ray waveband.

Another very good example is to look at the region closest to the Earth in which massive stars are being born now. In the optical waveband,

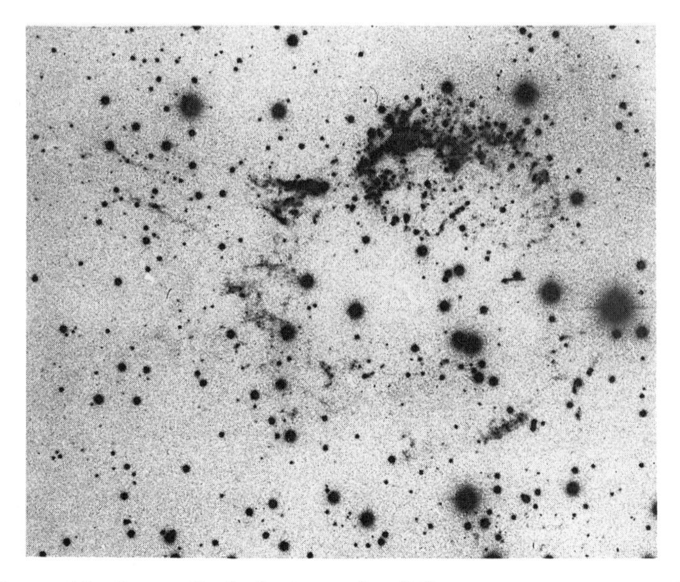

Figure 19. An optical photograph of the supernova remnant Cassiopaeia A. The wispy filaments are moving away from the site of the exploding star at a speed of about 10 000 kilometres per second.

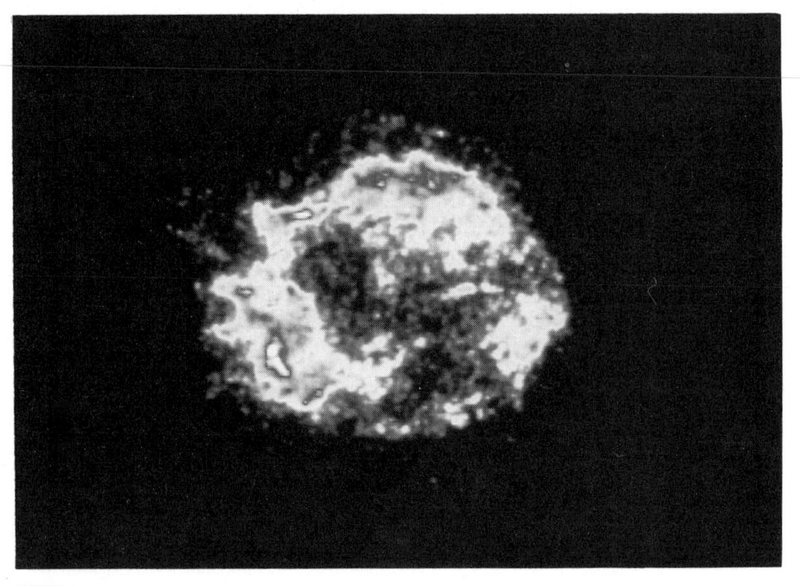

Figure 20. An X-ray image of the supernova remnant Cassiopaeia A. It is apparent that gas has been heated to very high temperatures by the explosion of the star.

we see the Orion Nebula with its beautiful optical filamentary structures (Figure 34). All this gas is at a temperature of about 10 000 K and is heated by the young blue stars, the four Trapezium stars, in the centre of the Nebula. When we look in the infrared and millimetre regions of the spectrum we see a completely different picture of very much cooler material out of which even younger stars are forming (Figure 37). We will have a great deal to say about this in the next chapter.

It is now time for us to look at the whole sky again but now we are to look at it in all the different wavebands mentioned above. First of all, let us remind ourselves of Figure 4 which shows the Universe as we know and love it, that is to say, in the optical waveband. One of the problems about astronomy is that the celestial sphere above us is laid out on the surface of a sphere but we have to draw our star maps on two-dimensional pieces of paper. The particular projection shown in Figure 4 is known as an Aithof projection and, in it, the plane of our Galaxy is placed right across the middle of the diagram with the centre of the Galaxy in the centre of the projection, the North Galactic Pole at the top and the South Galactic Pole at the bottom. Areas are preserved on the diagram but, to do this, you have to bend your coordinates rather badly and that is why the diagram has a funny squashed appearance towards the North and South Galactic Poles. This optical picture of the sky includes all of the stars down to about 10th magnitude as well as all the nebulae seen in the plane of the Milky Way. Most of the light we see in this picture is due to objects which are radiating at temperatures between about 3000 and 10 000 K and you can see that the principal objects are gas clouds and stars. There are very good astrophysical reasons why it is that stars radiate with surface temperatures which lie in this temperature range and we will have to go into this later. The galaxies are simply collections of stars and gas clouds and that is why we observe them so beautifully in the optical waveband. You can see our two dwarf neighbouring galaxies, the Magellanic Clouds, to the lower right of this picture.

Let us look at a hotter image of the sky. Figure 21 is a map of the sky in the X-ray waveband made by the HEAO-1 satellite observatory and you can see that the picture is totally different. There are some objects concentrated towards the Galactic Plane but they are far fewer than in the case of the optical picture and the reason is very simple. The very, very hot objects which radiate at temperatures of about 10 000 000 K are very rare objects in our Galaxy. There are two broad classes of Galactic object. Many of them are very special types of binary stars in

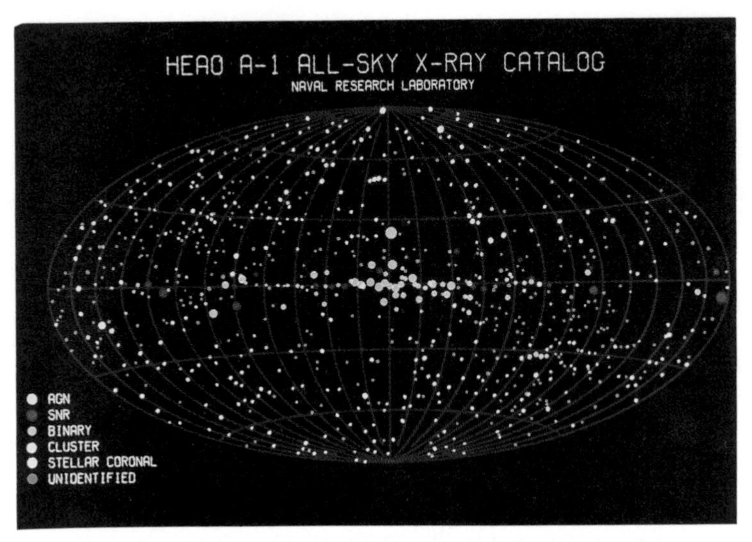

Figure 21. An X-ray image of the whole sky in the same projection as Figure 4 obtained from the HEAO-1 satellite.

which a very compact star radiates at a very high temperature because of the mass falling from the companion star onto it. The other broad class consists of the remnants of exploding stars (supernova remnants) and the dead stars formed in the explosions. We will learn much more about the binary X-ray sources when we discuss active galactic nuclei and quasars in Chapter 3. You will also see that there is a large population of objects away from the Galactic Plane, many more than there were in the optical picture. These are mostly active galaxies and quasars at very large distances. The reason they appear on the sky is simply because these objects are very luminous X-ray emitters. There are also a few clusters of galaxies which contain very hot gas indeed in the space between the galaxies.

We can now go to even higher temperatures and look at the γ-ray map of the sky (Figure 22). This image of the plane of our Galaxy was made by the European Space Agency's COS-B satellite. You will notice that it is not a complete picture of the sky because these γ-rays arrive at very, very low rates indeed. Over the period of 10 years when this satellite was in orbit only about 3000 γ-ray events were registered but the picture resulting from these is fascinating. You can see that the bulk of the radiation is confined to a thin layer in the plane of our Galaxy and the temperature of this gas must be well over one billion K. In fact,

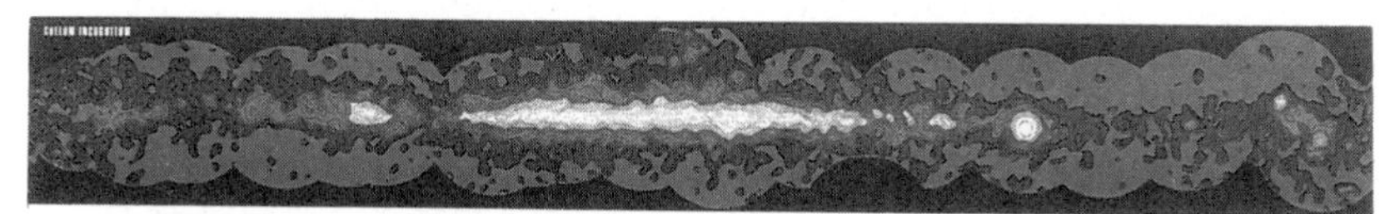

Figure 22. The γ-ray map of the Milky Way as observed by the COS-B satellite.

this radiation is associated with very high energy particles which fill the space between the stars. The γ-ray radiation is due to collisions between this very hot gas and the cooler matter lying between the stars.

Let us now go to wavelengths much longer than optical wavelengths, corresponding to about 60 or 100 μm. Figure 23 shows the map of the whole sky observed in this waveband by the IRAS satellite. These wavelengths correspond to temperatures very much less than room temperature - in fact, to about 30 to 50 K, corresponding to about -240 to -220 °C. You can see once again that there is a very narrow distribution of intense radiation lying along the plane of our Galaxy. This matter is very cold but the radiation is very intense. It is due to the radiation of cold dust lying in the plane of our Galaxy. Why is it at this low temperature? The reason is that, although it is not very hot, it still must be heated by something or else it would just cool down to a very low temperature indeed by radiating away its internal energy. What is happening is that, in the plane of the Galaxy, dust grains are heated by the hot radiation of stars in the process of formation or by stars which are dying. This is believed to be the source of most of this far infrared radiation and we will have a great deal to say about this in the next chapter when we study how stars form. Infrared radiation of cool dust is one of the most important processes for understanding how stars are formed. There is one very beautiful example of the importance of these observations. You will remember how the optical picture of the Galaxy (Figure 4) was obscured by dust. One key feature of the infrared waveband is that dust becomes transparent at these wavelengths. This is why we were able to obtain such an elegant image of the disc and bulge of our Galaxy from the near infrared observations made with the COBE satellite (Figure 5) - we really do live in a spiral galaxy with a central bulge!

Let us go cooler still. As we go cooler, we expect to observe colder and colder objects and Figure 24 is a picture of the whole sky at a wavelength

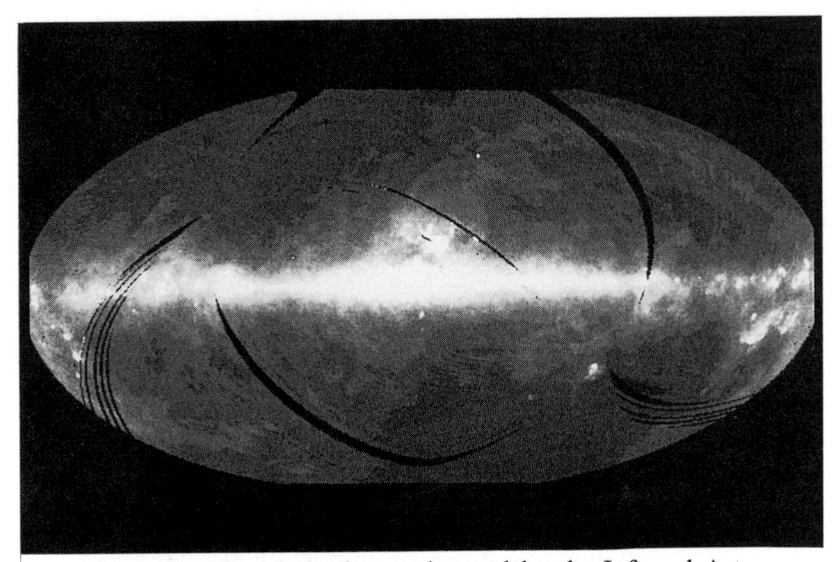

Figure 23. The whole sky as observed by the Infrared Astronomical Satellite IRAS in the wavebands 60 to 100 μm.

of only 5.7 mm. This corresponds to temperatures only about 1 K above absolute zero. This picture differs dramatically from all the others we have looked at so far. In the other pictures, we saw a clearly defined Milky Way and then there was sometimes some evidence for sources away from it. In Figure 24 you can see that there is a vestige of the Milky Way running along the centre of the diagram telling us that there is still some cool gas present in the plane of the Galaxy but you will notice that the picture is dominated by intense radiation which is hot in the top right of the map and cool in the bottom left. This picture is in fact not the total intensity of the radiation but simply deviations of the temperature away from a uniform temperature of 2.736 K. This is a picture of the famous microwave background radiation, the radiation which was created in the early stages of the Hot Big Bang, and which we will talk about in Chapter 5. The important thing about this cool radiation is that it has an essentially perfect black-body spectrum, one of the most remarkable observations of modern cosmology which was only made last year by the Cosmic Background Explorer (COBE) of NASA. I must show you this beautiful spectrum - this is most perfect black-body spectrum I know of anywhere in nature (Figure 25). We will explain why it has this form of spectrum in Chapter 5.

If we go to even longer wavelengths, radio wavelengths, we would

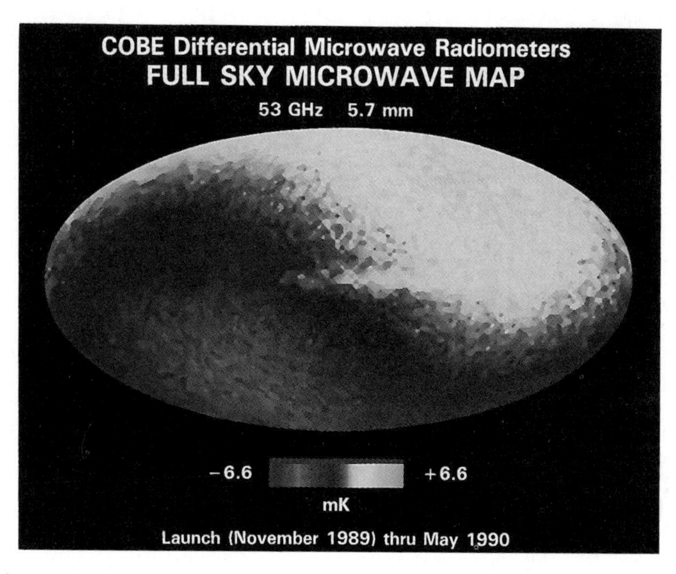

Figure 24. The map of the whole sky as observed in the millimetre waveband at a wavelength of about 5.7 mm by the COBE satellite. The image is dominated by the dipole component of the microwave background radiation.

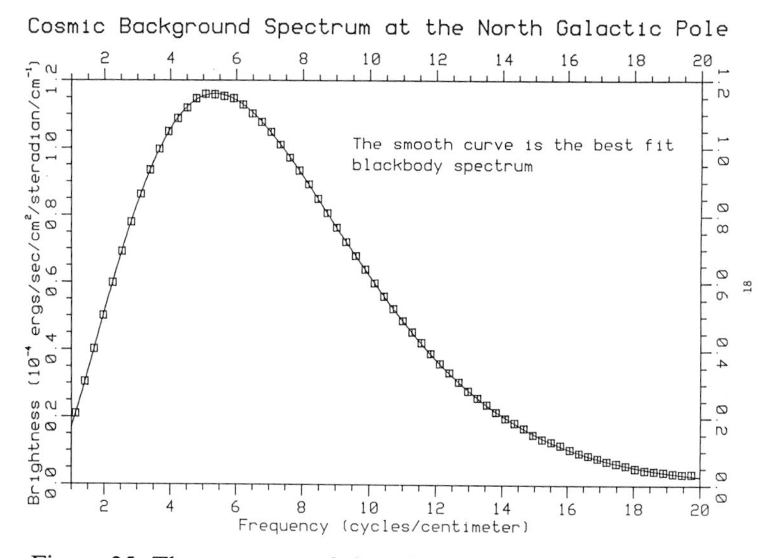

Figure 25. The spectrum of the microwave background radiation as measured by the COBE experiment. The spectrum is that of a perfect black body at a temperature of 2.736 K.

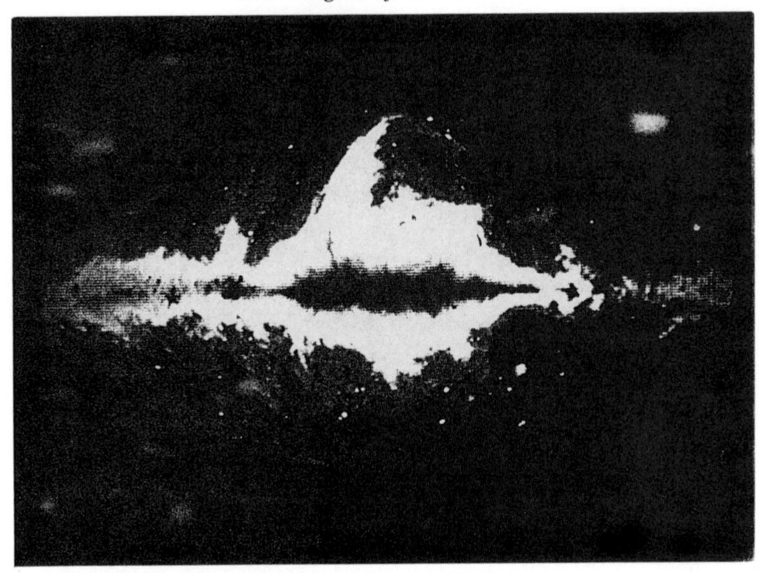

Figure 26. The radio emission from the Galaxy at a wavelength of 73 cm as measured by the group at the Max Planck Institute for Radio Astronomy at Bonn.

expect to see the coldest material of all but in fact, this is not the case. The picture we see in Figure 26 shows again the Galactic Plane but now there are spurs and plumes bubbling out of the plane. Ultra-high energy electrons gyrating in magnetic fields produce this intense emission which is observed in the radio waveband. This is similar to the radiation which is seen in many other active and exotic objects and we will have a great deal to say about this process in Chapter 3.

The important thing we have learned in this chapter is that we must look at the Universe in all wavebands in order to appreciate fully what it really looks like. We know that the Universe contains stars, galaxies and hot gas, but we also have to explain where the high energy particles and magnetic fields came from. Our modern view of the Universe is one in which there are many hot and cold phases present in most objects and what we have to do in the next four chapters is to put all these observations into context to create a convincing picture of the origin and evolution of stars, galaxies, quasars and the Universe itself.

2

The Birth of Stars and the Great Cosmic Cycle

The subject of this chapter is the stars and how they were formed. Most of the visible mass of our Galaxy is tied up in stars and consequently there must be rather efficient ways of condensing what started out as very diffuse gas into the objects we call stars. Let's look at common or garden stars first of all. How do they work? The answer was only fully appreciated once the process of nuclear fusion was understood. Before that time, calculations had been done to work out how long a star like our Sun could keep glowing at its present rate for the known age of our Solar System which is about 4.6 billion years. If you do that calculation, you convince yourself very quickly that the Sun cannot be powered by chemical energy or any of the more conventional forms of energy generation. In 1920, however, the British scientist Arthur Eddington noted in his address to the British Association meeting in Cardiff that the nuclear conversion of four hydrogen nuclei into a helium nucleus can release 1/120 of the rest mass energy of each hydrogen nucleus. This prescient conclusion turned out to be absolutely correct and this nuclear process is indeed the energy source of the Sun.

The process of nuclear energy generation in the Sun is very simple. Most of the matter in the Universe is in the form of hydrogen which is the lightest of all the chemical elements. Each atom of hydrogen consists of a nucleus, in this case a single proton with a single positive charge, surrounded by the negative charge cloud associated with a single electron. Heavier elements such as helium, carbon and oxygen have nuclei which are heavier than those of hydrogen and are composed of particles called

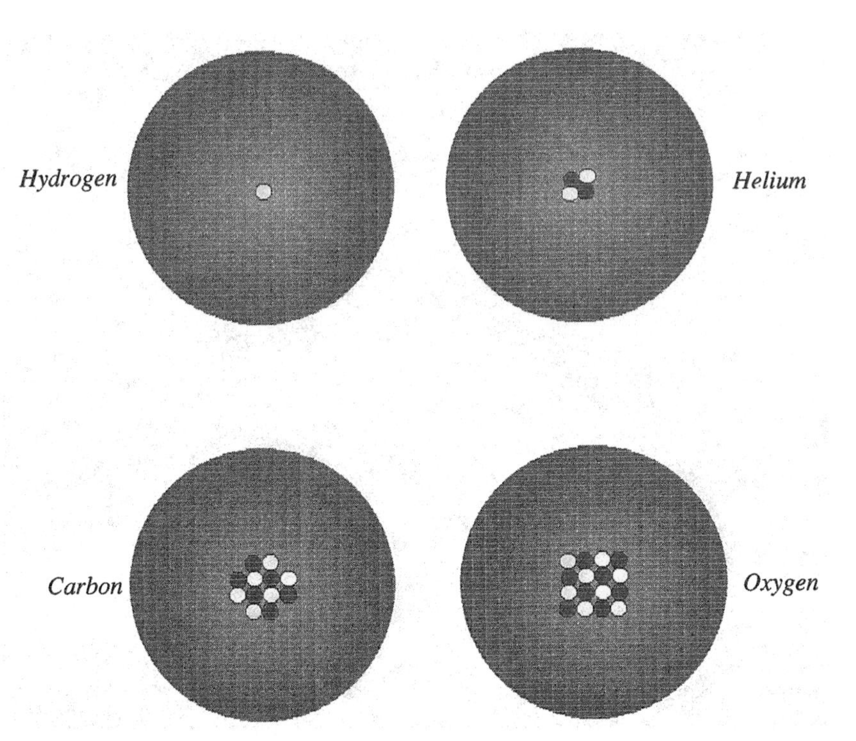

Figure 27. Sketches of the atoms of hydrogen, helium, carbon and oxygen. Hydrogen consists of a single proton which has a single positive charge surrounded by the charge cloud associated with a single electron which has negative charge. The helium nucleus consists of two protons and two neutrons. The neutron has no electric charge and so the charge cloud is associated with two negatively charged electrons. In the same way, the nucleus of carbon contains six protons and six neutrons and oxygen has eight protons and eight neutrons.

nucleons, which are either protons or their neutral twins, the neutrons (Figure 27).

The fact that the nucleons stick together in nuclei rather than remain isolated means that it is energetically favourable for the nucleons to combine together to form heavier nuclei. It is a general property of natural processes that they try to reach their lowest energy state. This means that, if we make the next heaviest element, helium, from hydrogen nuclei, then some energy is left over. If there wasn't energy liberated in this process, helium and the heavier elements could not be formed

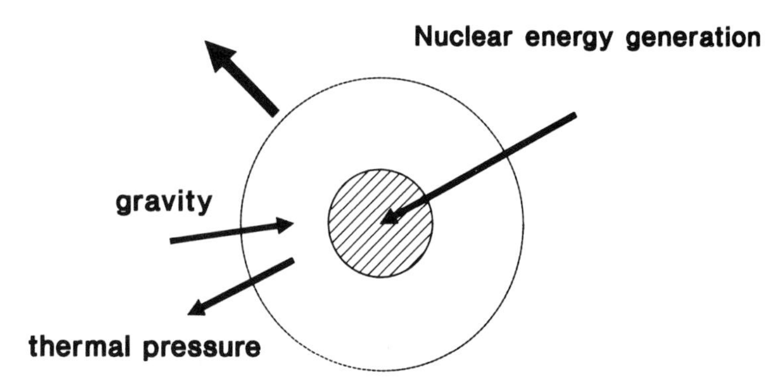

Figure 28. A schematic diagram showing the internal structure of the Sun. The Sun is held up by internal pressure gradients against the attractive force of gravity.

naturally out of hydrogen. It is this reaction - the synthesis of helium out of hydrogen - that powers stars like the Sun. The understanding of how stars like the Sun work is one of the great triumphs of modern astrophysics and is among the most precise of the astrophysical sciences. The region in which these nuclear reactions take place occupies roughly the central 10% of the Sun by radius. In order for the nuclear reactions to take place, the temperature in the centre must be very high indeed. Once the reactions get going, however, they generate a huge amount of energy in the central regions which then diffuses out through the Sun to its surface where we observe it as optical radiation. This is what we mean by normal stars - they are regions in the Universe in which the gas density and the temperature are so great that nuclear reactions in their interiors are initiated and are the source of their great luminosities.

Now this is all very well but how do we know it is correct? After all, we can only observe directly the surface of the Sun, the region from which all light comes, and we can only infer indirectly what the innards of the Sun must be like from its surface properties (Figure 28). This continues to be one of the most important aspects of solar physics but there are now two other ways of observing deep inside the Sun. The first of these is now more than 20 years old and is to look for particles created in the nuclear reactions which hardly interact with the solar material at all. These are the famous particles known as **neutrinos**.

What are neutrinos? They are particles involved in certain types of nuclear reactions, the weak interactions, and were predicted to exist in the 1930s. They are however very difficult to detect because they have hardly any interaction with matter. They were only detected experimentally in the 1950s by placing huge detectors underneath a nuclear reactor. These neutrinos are liberated in absolutely enormous numbers in the centre of the Sun. Because they interact so weakly with matter, they escape more or less directly from the centre of the Sun without colliding with any of its material. Each second there are roughly 10^{15} of them passing through each square metre at our distance from the Sun but they interact so weakly with matter that we do not recognise their presence.

It is very important to make special efforts to detect this flux of solar neutrinos because it can tell us whether or not we have our nuclear physics and astrophysics correct in the centre of the Sun. In fact, the solar neutrinos have been detected by an enormous effort over the last 20 years by Dr Raymond Davies and his colleagues in South Dakota. They built a huge underground tank containing the chemical perchloroethylene, similar to cleaning fluid, which has the great advantage that it contains lots of chlorine atoms. Very, very rarely, a neutrino hits a chlorine nucleus and converts it into a radioactive argon nucleus. The radioactive argon gas is swept out and the amount of argon produced is measured by the amount of radioactivity due to the decaying argon isotopes. The good news is that solar neutrinos have been detected - the bad news is that there are too few of them. This is one of the great problems of modern astronomy.

When calculations are made using the very best nuclear physics and astrophysics, there is still a discrepancy of about a factor of 3 between what is expected and what is observed (Figure 29). The solution to this problem is not known at the moment. We don't know whether there is some piece of nuclear physics which is not yet properly included in the calculations or whether the astrophysics which determine the properties of the Sun are not quite correct. Further crucial information will be available quite soon because, so far, the experimenters have only been able to search for relatively high energy neutrinos and there are in addition lower energy neutrinos which are produced in much greater numbers. Two important experiments using gallium as a detector are currently underway to detect these lower energy neutrinos and the first results should be available soon. These experiments will determine whether or not there is a real difference between the predictions of the

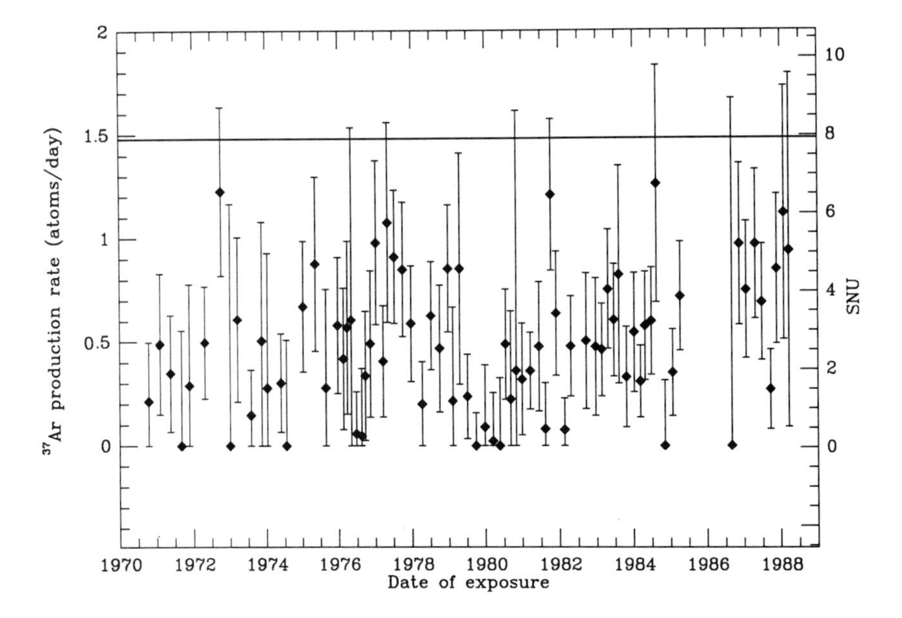

Figure 29. A comparison of the observed flux of solar neutrinos with observations carried out at the Homestake Goldmine through the period 1970 to 1988. It can be seen that the average production rate of radioactive argon, 0.462 atoms of ^{37}Ar per day, is significantly smaller than the expected production rate of about 1.5 atoms per day.

models and the theory. So, keep your eyes on the scientific press - the results are crucial for the whole of astronomy.

A second remarkable development of the last 10 years has been a completely new way of studying the internal structure of the Sun. If you hit a bell, it makes a ringing noise and, if you analyse the frequencies of vibration of the bell, you can work out something about the structure of the bell and why it makes the sound it does. In exactly the same way, if you kick the Sun, it resonates at its natural frequencies of vibration. The Sun is in fact being constantly perturbed from the inside by the turbulent and convective motions in its interior but the amplitude of the resulting vibrations of the Sun are very small and it requires extreme experimental sophistication to detect them at all. These have, however, now been measured and you can understand why this is such an exciting area of research. The frequencies of oscillation of a bell, or the Sun, depend not only upon its surface properties but also upon the bulk properties

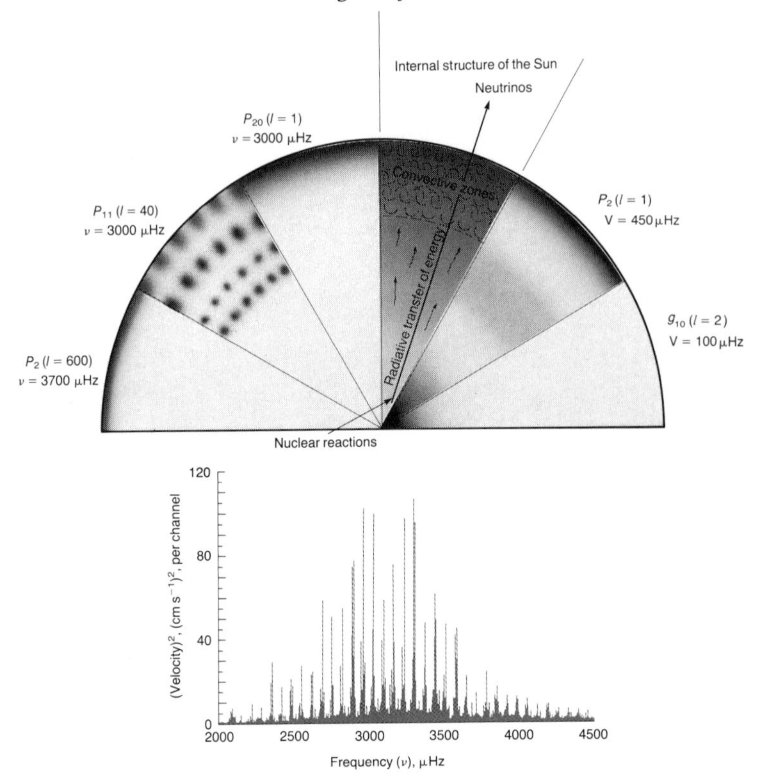

Figure 30. (top) A schematic diagram showing the internal structure of the Sun. Also shown are examples of how different modes of oscillation of the Sun can be used to probe the conditions of temperature and density at different depths in the Sun. (below) An example of the frequency spectrum of solar oscillations showing the fine structure splitting associated with coupling between rotation and normal modes of oscillation of the Sun.

of the object, all the way through to its centre. Therefore, by studying the resonances of the Sun, we can infer what its internal density and temperature structure must be (Figure 30).

The theorists can use these oscillations to work out, not only the temperature and density structure but also the internal rotation of the Sun. An example of how well the internal structure of the Sun can be determined by this technique is shown in Figure 31. It can be seen that the agreement between theory and observation is remarkably good. In some of the most recent experiments by a research group at Birmingham University, it has been shown that the astrophysical models

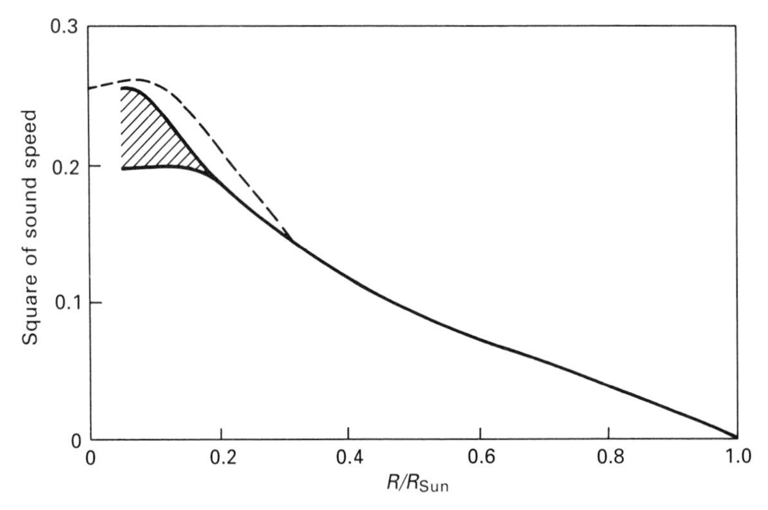

Figure 31. The speed of sound inside the Sun as a function of distance from its centre as derived from theory (dashed line) and the limits obtained from solar seismology (solid lines).

for the structure of the Sun seem to be remarkably good right into the core of the Sun. These results have been obtained from the first sets of systematic observations of solar oscillations. The next generation of instruments both on the ground and in space will greatly improve our understanding of the internal structure of the Sun. One of the exciting projects which astronomers are already attempting is to observe these same types of oscillations in nearby stars. It must be emphasised, however, that stellar seismology on even nearby stars is very difficult and will require very large telescopes and lots of observing time.

Now, let's put together a picture of how the stars evolve. This is what I call the **great cosmic cycle**. Stars like the Sun are very stable in the sense that they burn the nuclear fuel in their centres steadily and the energy released diffuses out through the outer regions, ending up as the light that we see through its surface. Most of the stars we observe in the Universe are in a similar state to the Sun in that their main source of nuclear energy is the conversion of hydrogen into helium. This is by far the longest phase of the star's lifetime and stars like the Sun, for example, remain in this state for about 10 billion years. All the stars in this state have a very well defined relation between their luminosities

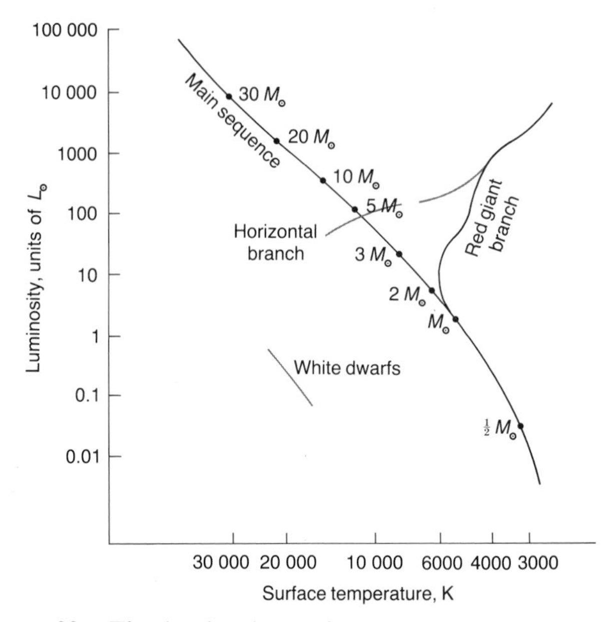

Figure 32. The luminosity-surface temperature diagram for stars, also known as the Hertzsprung-Russell diagram. Stars do not occupy all regions of this diagram but form various sequences. Most stars lie along a band which stretches from the bottom right to top left of the diagram which is known as the **main sequence**. The **giant branch** extends from the main sequence to the top right of the diagram. The masses of stars at various points along the main sequence are indicated relative to the solar mass $M_\odot$.

and their surface temperatures - the stars which follow this relation are referred to as **main sequence stars** (Figure 32).

During this long period on the main sequence, the star continues to burn hydrogen into helium but then, when about 12% of its mass is converted into helium, the star becomes unstable. The central regions of the star begin to contract while the outer layers expand. This process results in the formation of what are known as **giant stars**. The centre keeps on contracting and heating up while the envelope expands by an enormous factor to create a star which is many thousands of times greater in radius than the main sequence star from which it formed. Because the giant stars are so big, they are very luminous and so, although they are much rarer than main sequence stars, they are not difficult to find. While this is happening, the core of the star continues to contract and heat up, as it tries to attain a more highly bound state. As the core contracts, it becomes hotter and eventually it is hot enough to convert the helium into carbon. As each new phase of nuclear burning is switched on, the star undergoes considerable internal rearrangement of its structure. Eventually, in the more massive stars, the carbon may be burned into yet heavier elements such as silicon and iron.

While these different nuclear burning processes take place, the star moves progressively up the giant branch shown in Figure 32 with various brief excursions across the diagram as the different nuclear burning processes are switched on. Eventually, the star runs out of possible stable states and the central regions collapse catastrophically. In that process, it is likely that an enormous explosion takes place and the star ends up as some sort of dead star - either as a **white dwarf**, a **neutron star** or a **black hole**. These objects are the end points of stellar evolution and we are going to have a lot to say about them in the next chapter. All I need say at this stage is that, when the star dies, it simultaneously ejects a substantial fraction of its mass back into the space between the stars - what we call the **interstellar medium**. The beautiful thing about this process is that, in the process of stellar evolution, hydrogen is converted into heavier elements, the star evolves onto the giant branch and then the processed elements are returned to interstellar space where they become the material out of which the next generation of stars can be formed.

We will demonstrate in a moment that stars form in dense regions of the interstellar gas. To illustrate this cosmic cycle - what I call the **great cosmic cycle** - Figure 33 shows the complete evolution of a star from its birth in a dense dust cloud through to its death as some form of dead star. This picture is, in fact, a time lapse picture, similar to those which are used to show plants growing and flowering in a matter of seconds. In Figure 33, the shutter of a camera has been opened about once every few million years. You can see what happens. The star forms very rapidly out of the interstellar gas and then settles down to a very long period as a main sequence star like our Sun. Our own Sun is about half-way through its life as a main sequence star, being about 4.6 billion years old out of a total lifetime of about 10 billion years, so we don't have to worry about running out of solar energy for some little time. You can also see that the death-throes of the star when its innards collapse and the outer envelope expands take place very rapidly compared with the main sequence lifetime of the star.

Just to complete our story of the properties of ordinary stars, why is it that they seem to come in quite a narrow range of masses? There

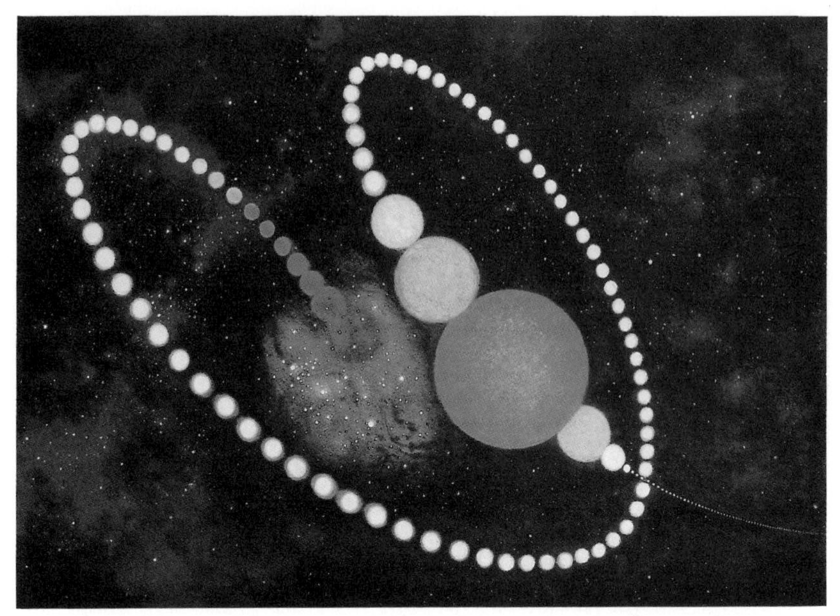

Figure 33. A "time-lapse" picture of the evolution of a star from its formation in a giant molecular cloud through the main part of its lifetime as a main sequence star and ending in its expansion to a red giant star and its subsequent demise as some form of dead star.

are good astrophysical reasons for this. The temperatures shown on the horizontal axis of Figure 32 are surface temperatures and, of course, the temperatures in the centres of the stars are very much greater than these, of the order to 10 million degrees or greater so that the nuclear reactions can be sustained. It is roughly correct that the surface temperature is proportional to the central temperature of the star. Now, the lower mass stars towards the bottom right of the main sequence have much lower surface temperatures than the Sun and correspondingly smaller central temperatures. Therefore, if the star is of too low mass, it will not be hot enough in its centre to maintain the nuclear reactions which convert hydrogen into helium. Theoretically, it is expected that stars with mass less than about 1/15 of the mass of the Sun will never be able to burn hydrogen into helium because the temperature is never high enough. It is expected that such objects will be rather inert bodies, not unlike large planets or huge lumps of rock. Collectively, this class of inert stars with masses less than about 0.06 times the mass of the Sun is known as the

brown dwarfs. One of the major unsolved problems in astronomy is how much mass in galaxies and in the Universe in general is tied up in these brown dwarfs. There could in principle be a great deal of it but it is very difficult to detect because it is cold and non-luminous.

At the high mass end, stars with mass about 50 times the mass of the Sun and greater are unstable. These are enormously luminous stars (see Figure 32) in which much of the internal pressure is provided by the pressure of radiation. These very massive stars are also so luminous that they burn up their nuclear fuel very rapidly. Thus, the combination of short lifetime and instability sets an upper mass limit to the stars. These astrophysical reasons explain why it is that most of the stars we observe in the Universe have masses in the range about 0.1 to 50 times the mass of the Sun.

Now, one of the big problems of astronomy and cosmology is how whole galaxies of stars and the large scale structure of the Universe came about. You can appreciate that we cannot really work out how the galaxies formed and evolved unless we understand how the stars themselves were born, how they live and how they die. Once we start making stars in a forming galaxy that influences the ability of the galaxy to make more stars because star formation reduces the amount of gas available to make more stars. However, once the first generation of stars has formed and evolved to the end of their lifetimes, they return processed gas enriched with heavy elements to the interstellar medium and that makes it easier for the gas to cool and so create the next generation of stars. Therefore, the problem of understanding what determines the rate of star formation is intimately tied up with the problems of understanding the origin and evolution of galaxies themselves.

One of the great problems of modern astronomy is disentangling how the stars are formed. We have learned a great deal about this problem over the last 10 to 15 years thanks to our ability to study the very earliest stages of star formation in completely new ways and I will devote the rest of this chapter to giving you some idea of how it is we can study these early stages of star formation.

There is one obvious problem which you can appreciate from Figure 33 and that is that the process of star formation takes place very rapidly compared with the lifetimes of the stars and therefore we have to catch the objects just as they are forming. One of the best ways of finding the very youngest stars is to search in regions which are known to contain very young stars. The most nearby nursery for young massive stars to the Earth is the Orion Nebula. Orion is one of the most famous

Figure 34. The Orion Nebula is optically the most conspicuous part of a huge region of star formation in the constellation of Orion. The clouds of hot gas seen in this picture are excited by four young blue stars, known as the Trapezium stars, which lie within the brightest part of the Nebula.

constellations in the sky and, if you look halfway down Orion's sword, you will see a rather diffuse patch of light. On a photograph taken with a large telescope, this is one of the most beautiful objects in astronomy (Figure 34). From the astrophysical point of view, this region is of special interest because it is a region in which we know massive stars are forming now. Some of these young stars are easy to identify and they are the very bright blue stars in the centre of the Orion Nebula. Four bright stars, known as the Trapezium stars, in the brightest part of the nebula are responsible for producing all the beautiful filamentary structure which surrounds the region. These Trapezium stars are young in the sense that they are no more than a million years old and therefore at a relatively early stage in their evolution. However, they have already made it - they know how to form! What we want to understand are the processes by which the nuclear reactions in their interiors were switched on.

There is one very obvious problem with this picture and that is that whilst it is very beautiful, you will also notice that it is very dusty. In fact, dust is present everywhere in the vicinity of the Orion Nebula and that has the effect of preventing us seeing deep inside the regions where the stars are forming. Let us have another look at the problem of dust. We know that there is a remarkable amount of dust in the plane of our own Galaxy and it prevents us seeing to large distances through the plane of the Milky Way (see Figure 4). We observe these large dust clouds in external galaxies as well (see Figure 6) and they can be a bit of a nuisance for many types of astronomical study. They often obscure interesting regions which we need to study - for example, the centre of our own Galaxy. What is happening physically is that the dust grains absorb and scatter the radiation incident upon them. From studies of the properties of the dust, we know that the obscuration in the optical waveband is caused by small solid particles about 1 μm in diameter, roughly the size of the particles of cigarette smoke. Carbon is one of the main constituents of the dust grains but there is also silicon present in some grains.

Fortunately, there is a very nice way of seeing through the dust and that is to observe the regions at rather longer wavelengths ($\lambda > 1\ \mu$m) than the optical waveband. At these longer wavelengths, the wavelength of the radiation becomes greater than the size of the grains and then radiation is hardly impeded at all. It is only when the grains are larger than the wavelength of the radiation that they absorb strongly. The best way of demonstrating this is by making observations of the Orion Nebula in the optical and infrared regions of the spectrum.

Infrared astronomy has been very difficult until recently because we have not had available the equivalent of photographic plates with which to take pictures of the sky. Through some wonderful technological developments, we have had access over the last 4 years to infrared electronic array detectors of very, very high sensitivity, which have enabled us to take pictures in the infrared waveband for the first time. Figure 35 is a photograph of the UK Infrared Telescope in Hawaii and you can see behind the primary mirror a really magnificent infrared camera built by my colleagues at the Royal Observatory in Edinburgh which is designed as a versatile camera for all sorts of astronomical study.

Let us see what happens when we look at the Orion Nebula in this way. Figure 36 is an optical picture of the central regions of the Orion Nebula showing clearly the four Trapezium stars. This is an optical photograph taken at a wavelength of about 0.5 μm. Now let's have a look at the infrared image made at wavelengths which are about a factor of 3 or 4 longer than this wavelength. Figure 37 is a composite photograph of

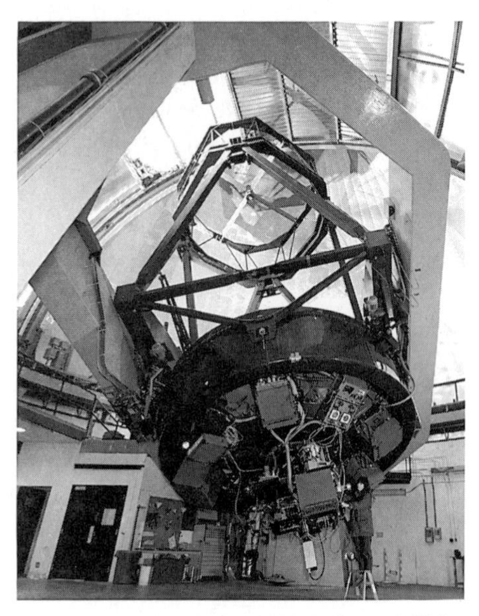

Figure 35. A photograph of the UK Infrared Telescope which is located on the summit of Mauna Kea in Hawaii. It is the world's largest dedicated infrared telesope. Behind the primary mirror can be seen the equipment associated with an infrared camera known as IRCAM built at the Royal Observatory, Edinburgh. This camera took the infrared images shown in Figure 37.

images made at 1.2, 1.65 and 2.2 μm. The effect is just like removing a great veil from in front of the region. The patchy dust obscuration has disappeared and it is now clear that there is indeed a cluster of stars present, roughly centred on the bright central stars. The four bright blue Trapezium stars are still there but, in addition, look at all the new objects which appear. These are totally absent from the optical image. There is plainly a great deal of activity going on. The really interesting regions are to the north of the Trapezium stars where you see enormously luminous infrared sources appearing. The two brightest objects are known as the Becklin-Neugebauer object (to the north) and the Kleinmann-Low Nebula (to the south) but the important thing is that these are very, very dusty regions indeed in which there is an enormous amount of infrared radiation escaping from embedded sources. The regions to the north of the Trapezium stars have luminosities more than 100 000 times the

Figure 36. An optical image of the central regions of the Orion Nebula, showing the Trapezium stars and the patchy obscuration by dust.

luminosity of the Sun but that radiation is all coming out as infrared radiation.

Now, why is the radiation so intense in the infrared waveband? The answer is very simple - the dust now acts as our friend, rather than our enemy. There are very, very young stars indeed, possibly even stars in the process of formation, what are known as **protostars**, buried deep inside the dense dust clouds in the Orion Nebula. Intense optical and ultraviolet radiation attempts to escape from these luminous objects but these radiations are strongly absorbed by the dust clouds in which they are embedded. In the process of absorbing the radiation, the dust is heated up and this heat is then re-radiated at the temperature to which the dust is heated. It is therefore important to work out the temperature to which the dust grains are heated close to young stars. It turns out that the temperatures are typically about 100-1000 K and therefore we can use our relationship between temperature and wavelength (Figure 16) to work out at what wavelength most of the radiation will be emitted. You will already have guessed the answer correctly - the radiation is emitted at wavelenths from about 3 to 30 μm, corresponding to infrared to far infrared wavebands at which the dust is itself transparent to radiation.

Figure 37. A composite infrared image of the Orion Nebula showing regions which are only apparent in the infrared region of the spectrum.

What is happening, in a nutshell, is that the dust is acting as a transformer which enables the young star to get rid of its radiation in the infrared waveband where it can readily escape from its dusty birth-place.

This process puts a completely new complexion upon the beautiful far infrared map of the Galaxy (Figure 23) which we discussed in the first chapter. You will remember that when we looked at the infrared map of the Galaxy, I said that it was the radiation of dust heated by internal sources of energy in dust clouds. The regions of star formation are among the contributors to that far infrared picture of our Galaxy.

This is very interesting but what does it tell us about the way in which stars form? The problem is how we can start off with very, very diffuse gas, such as is found in the space between the stars, and then convert it into stars which have densities which are about 10^{24} times denser than the interstellar gas. Fortunately, we know that the interstellar gas is rather unstable. It does not just sit there and do nothing because there are all sorts of influences acting upon the gas which do not allow it to remain in a quiescent state. This leads us to the study of the behaviour of gas clouds under gravity.

If we have an isolated cloud, then we can show that, if it is massive

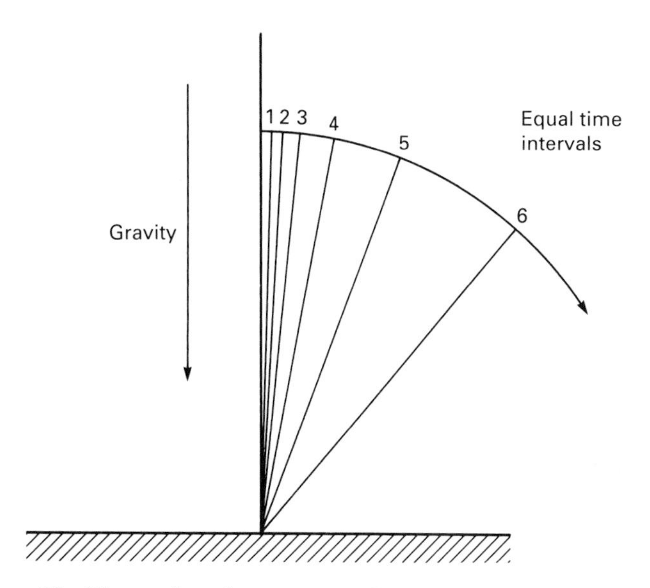

Figure 38. Illustrating the process of exponential growth under gravity by a falling pointer. The angle from the vertical increases by the same fraction in equal time intervals.

enough, it will collapse under its own gravity. What this means is that, so long as there isn't too much turbulence or pressure inside the cloud, it will collapse and what is particularly pleasing is that the collapse develops exponentially. In other words, it will halve its radius in equal time intervals so that we find a run-away situation in which the density can grow very rapidly to very high values. This is the famous instability discovered by Sir James Jeans in 1902 and is appropriately known as the **Jeans instability**.

There is a very simple way of demonstrating this form of collapse under gravity. Suppose I take a pointer and stand it on end. If I let it go, you all know that it will fall over, even if I do not push it. Suppose we measure the angle from the vertical direction in equal time intervals. You will find that the angle increases by the same proportion in equal time intervals as illustrated in Figure 38. This is what we call **exponential growth** and it has the excellent property that we can start with infinitesimally small perturbations and still the pointer (or star) collapses under gravity.

Now, if gas clouds in the Galaxy are big enough, they collapse in the same way and so we have good reason to believe that we at least know how the collapse of clouds can take place. There are other ways in which the collapse can get started and it's not at all clear which of these is the most important. For example, we know that there are spiral arms in spiral galaxies, which contain lots of dust and gas, and these can strongly perturb the distribution of gas. The spiral arms can help the gas to clump to quite high densities and the Jeans instability can then take over. We can also create rather dense gas if a diffuse gas cloud is hit by the expanding shell of an exploding star. The result is that we believe there are a number of ways in which we can enhance the average density of the interstellar gas to significantly higher densities.

Can we see these clouds? The answer is "Yes, we can!" As the clouds increase in density, they cool by radiation. In a typical large cloud, we expect there to be dust present and the combination of the dust and gas are very effective in keeping the gas very cold indeed by exactly the same mechanism by which the protostars lose energy - i.e. by the radiation of heated dust. In fact, these large clouds cool to such low temperatures - about 10-30 K - that we expect the gas to be in molecular rather than atomic form inside the dust clouds and we can search for the molecular line radiation of these molecules in the millimetre and submillimetre waveband (refer again to Figure 16). One of the great realisations of modern astronomy has been that the plane of our Galaxy is full of these **Giant Molecular Clouds**. In fact, the giant molecular clouds in the constellation of Orion are as big as the constellation itself. If our eyes were sensitive to millimetre waves rather than to optical radiation, the giant molecular clouds which we observe through rotational emission lines of molecular species such as carbon monoxide would be by far the most striking objects in the sky. We now know that there is as much gas in the disc of the Galaxy in the form of molecules as there is in neutral hydrogen and that the molecules are mostly confined to giant molecular clouds such as that present in the constellation of Orion.

We can now refine the problems to be solved quite a bit. How is it that once the interstellar gas is clumped into these cold giant molecular clouds that we can form stars? This is not a solved problem by any means, but we have a number of important clues as to what has to happen. There are, in fact, three big problems to be overcome. The first is the fact that, in order to make stars at all, the gas cloud has to lose the energy which it gains when the cloud collapses. If we simply take a cloud and let it collapse, then the kinetic energy of collapse heats up the interior of the cloud until the pressure of the hot gas stops the collapse. So, we have to find effective mechanisms by which, when the star is collapsing, it is able to radiate away efficiently all the energy generated by the heat of collapse. This is where, again, the dust plays a vital role, because whilst it prevents us seeing inside these dark regions, it is quite extraordinarily effective in getting rid of heat. As soon as heat is generated inside the cloud by collapse, its radiation is absorbed by the little dust grains which transform it to much longer wavelengths at which the radiation can escape from the clouds. So, we should be jolly glad that there is so much dust around. It may prevent us seeing inside these dense regions but, without it, the stars would find it much more difficult to condense to very high densities from diffuse interstellar matter. This is almost certainly the solution to the energy problem and it explains why regions of star formation are such powerful sources of infrared radiation.

Although we may be able to get rid of the energy, we have to think about the dynamics of the collapsing cloud as well. One of the big problems is that there is a great deal of turbulence in the regions where stars form and therefore it is very likely that any little bit of cloud we select is going to have a certain amount of rotation associated with it. This produces a big problem. If the cloud is rotating and it begins to collapse, we find that the rotation is amplified. This is exactly the same process by which an ice skater can spin so rapidly. The skater rotates slowly with arms outstretched and then, by drawing the arms in, speeds up enormously. Even more spectacularly, you can take a weight in each hand and sit on a rotating stool with your arms outstretched. Once your friends make your stool rotate and you pull in your arms, you speed up by a big factor. If you try this experiment, please wear a seat belt! Newton's laws of motion won't stop you falling off if you are unbalanced!

Now, exactly the same process can occur if you try to make stars collapse. Any rotation gets vastly exaggerated because of what we call the **law of conservation of angular momentum**. We've got to find an efficient way of making sure that any collapsing object can get rid of its rotation which can stop the process of collapse. If you try standing upright on a rotating roundabout you can feel a force pushing you outwards - what is called the **centrifugal force**. The rotation can become so great that the centrifugal force stops the collapse and prevents the star forming.

There is another problem which is rather similar and that is that there is very likely to be a weak magnetic field present in the collapsing cloud. Even if there is a very small number of free electrons in the cloud, that is quite sufficient to tie the magnetic field to the gas in the cloud. Now, we know that these free electrons are present in the clouds, probably

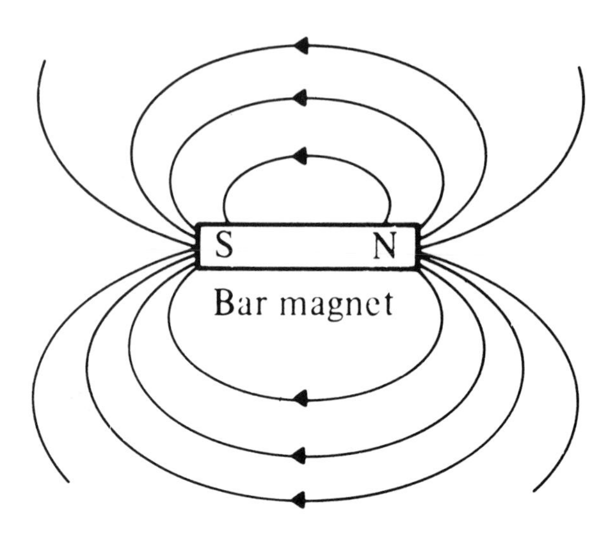

Figure 39. Illustrating Faraday's concept of magnetic field lines about a magnetic dipole. The field strength is proportional to the number of field lines per unit area perpendicular to the lines of force.

produced by collisions between very high energy particles and atoms and molecules in the clouds. There is a very elegant theorem which tells us that we can understand how the magnetic field behaves if we think of the magnetic field lines as being frozen into the gas. It was the great Michael Faraday who demonstrated here at the Royal Institution that you can always understand the behaviour of magnetic fields in terms of the properties of magnetic lines of force. Figure 39 shows the magnetic lines of force about a magnetic dipole, similar to the Earth's dipolar magnetic field. According to Faraday, the strength of the magnetic field is just proportional to the number of field lines passing through unit area and so the field is strongest at the poles and weakest at the equator.

Now, there is a big problem associated with these magnetic fields which is illustrated by the diagrams shown in Figure 40. Suppose the magnetic field lines are frozen into a rotating disc of gas and the disc does not rotate like a solid body, but say the velocity of rotation of each annulus of the disc rotates at the same speed. Then, as the disc rotates, the field lines get stretched in the direction of rotation. You see what

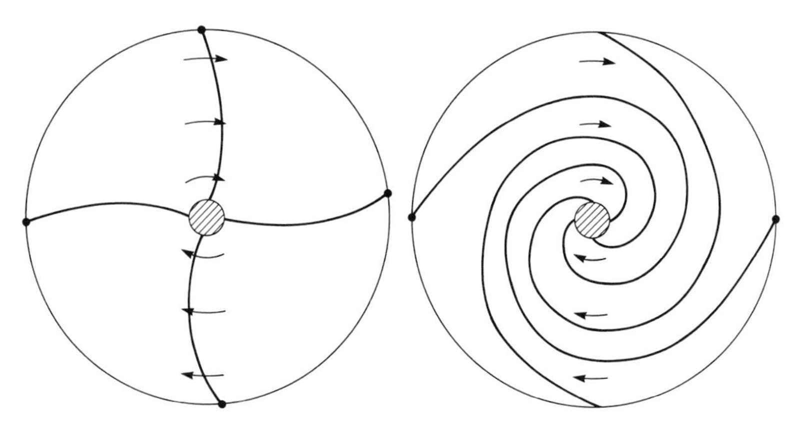

Figure 40. Illustrating the amplification of magnetic fields under magnetic flux freezing.

happens - Faraday told us that the strength of the magnetic field is just proportional to the number of field lines per unit area. As you start to wind up the magnetic field, the number of field lines per unit area increases and, therefore, if the magnetic field is frozen into the gas, the magnetic field is going to increase in strength inexorably under the effect of the rotation of the gas cloud. That is also rather bad news because it means that you could eventually build up so much magnetic energy that the collapse could be stopped.

There is no agreed solution to the last two problems, but there is one rather pretty idea which I think may be along the correct lines. One of the remarkable things we see among these very young stars are jets of molecular material which seem to be blown out of the poles of the parent objects. Figure 41 shows millimetre images of two of these types of source which are called bipolar outflow objects. Very much to our surprise, we find that the material is not collapsing onto the centre of the star but is actually being expelled in jets from a compact region where a star is forming. One side of the jet seems to be coming towards us and the other going away from us. A schematic picture of what is found to occur in these sources is shown in Figure 42. One picture of what is

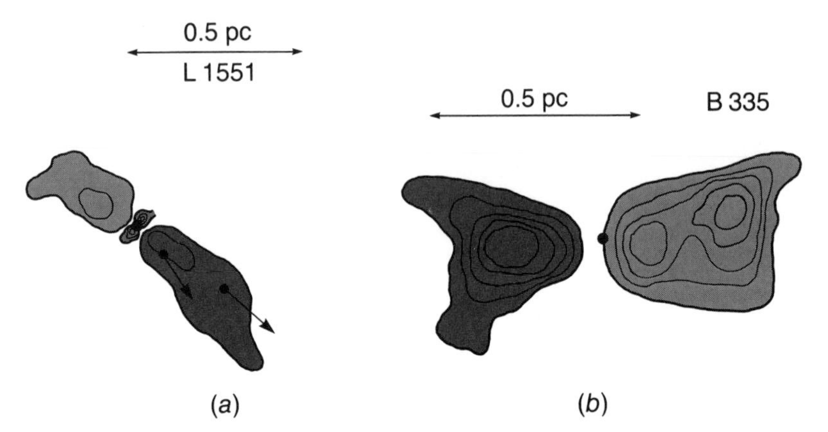

Figure 41. Two examples of high velocity bipolar outflows associated with very young stars. The latter are indicated by black dots. The darker lobe is moving away from the observer whilst the light one approaches the observer. (a) The bipolar outflow in the source L1551. In this case there is also evidence for a molecular disc of denser gas perpendicular to the direction of the outflow. (b) The source B335.

going on is that the rotation winds up the magnetic field lines in such a way that the field gets so strong in these dense regions that the magnetic field lines simply reconnect and all the stress in the magnetic field goes into ejecting clouds of material out along the axis of rotation (Figure 43). That would be a marvellous way of solving two of the problems at once. We are able to get rid of the magnetic field - it's just gone twang when the lines of force reconnect - and we've also got rid of the rotation as well by expelling it along with the ejected material.

You will notice that we are not only creating a preferred axis along which the material is ejected but also that there is a natural plane perpendicular to the rotation axis in which a rotating disc of dust and gas can form. Evidence of such dense discs of material about the axis of the outflow can be seen in the case of L1551 (Figure 41(a)). Recently, these dusty molecular discs have been observed by their thermal dust radiation. These observations were made in Hawaii with the James Clerk

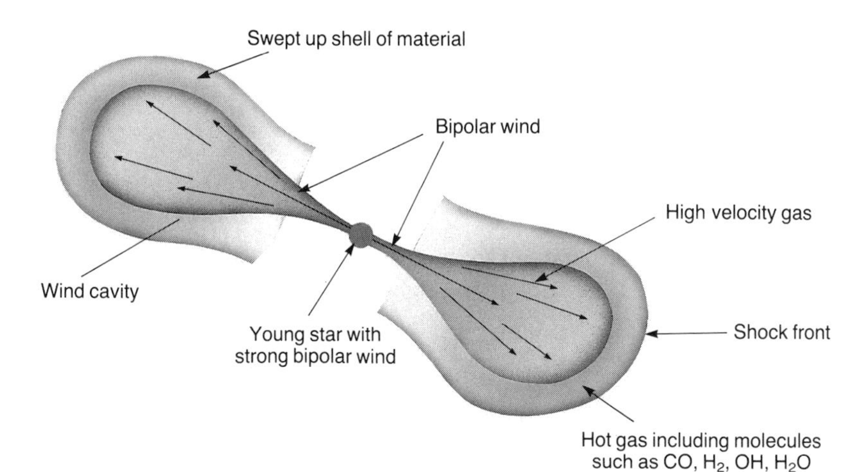

Figure 42. A model of the bipolar outflow of matter from a newly forming star. The outflow is supersonic and compresses the surrounding molecular gas. The heating of the molecular gas by the outflow and the cooling by molecular line emission results in a temperature of about 2000 K at which temperature it can be observed through its infrared molecular line emission.

Maxwell Telescope and in Figure 44 you can see evidence of a dusty disc about a young object in the system LkHα234. It is not clear yet exactly what the astrophysical significance of these discs is but we know that there must be both dust and molecules present in these regions. Similar phenomena have been seen by the IRAS satellite and it appears that these cool highly flattened discs of material are common features of many bright stars. One of the big questions is whether or not these dusty discs are indeed the earliest stages of the formation of planetary systems about young stars. What many of us are hoping is that we will be able to understand the process of planet formation by direct observations of these very young, flattened discs about young stars.

There is one other very intriguing thing about these discs about the young stars and that is the presence of molecules and, of course, molecules are needed in order to begin the process of forming biotic systems. A biotic system is just a rather fancy word for a molecular soup in which

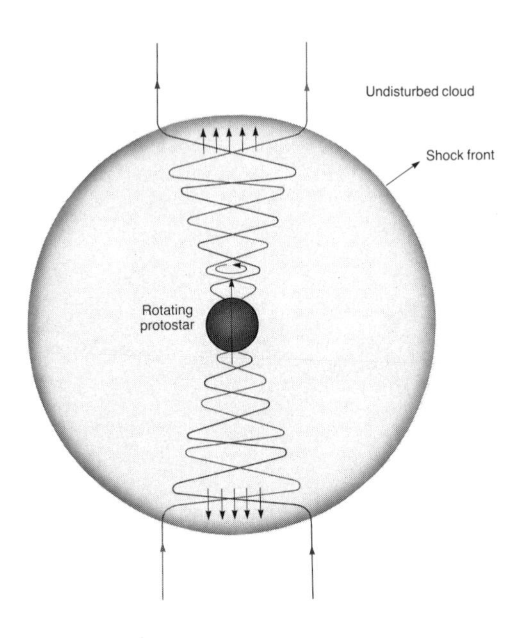

Figure 43. A sketch of a possible mechanism for the formation of bipolar outflows in protostellar objects. As the material collapses onto the protostellar object, the magnetic field is wound up and the magnetic field energy density increases. It is thought that the large magnetic pressure causes a shock wave to move out through the infalling plasma. The magnetic pressure increases, the winding-up continues, and eventually the pressure is released by the magnetic field and the plasma tied to it being ejected along the poles of the rotating object.

biochemical molecules necessary for the the creation of primitive life forms might be synthesised. One of the great discoveries of millimetre astronomy has been that the Galaxy is just crammed full of molecules. Over 60 different molecular species have now been identified in the interstellar gas, ranging from simple molecules like molecular hydrogen, carbon monoxide and the hydroxyl radical to complex molecules such as ethanol, long acetylenic chains, ring molecules such as H_3^+, SiC_2 and C_3H_2 and so on. Many complex molecules are believed to be created by surface reactions on the dust grains. Some molecular species have been found in interstellar space which are difficult to observe in the laboratory because they are very reactive and hence have very short lifetimes under laboratory conditions. They can, however, survive in the very low density conditions of interstellar space. The complete list of known molecules

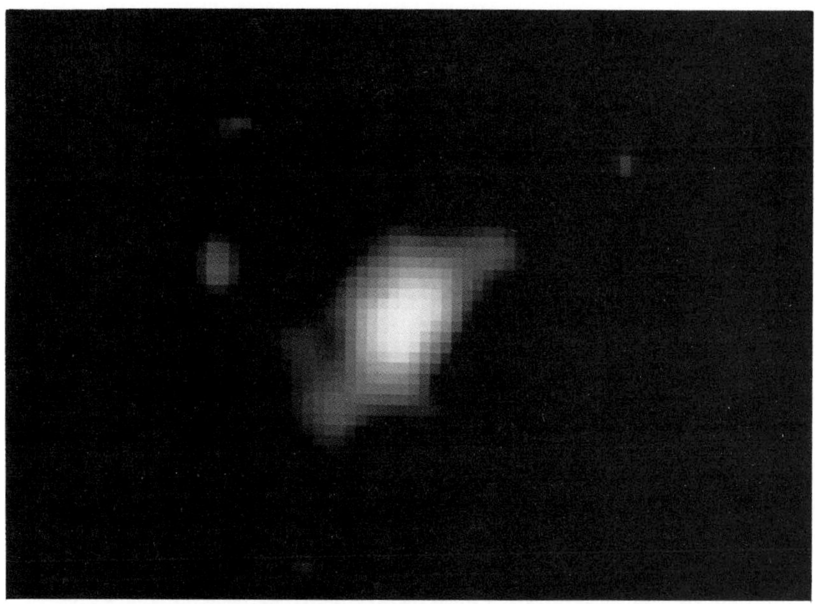

Figure 44. Submillimetre observations of a dusty disc about a young stellar object in the system LkHα 234 from observations made with the James Clerk Maxwell Telescope.

includes everything which is believed to be essential to start the process of biological synthesis. Interestingly, no amino-acid has yet been found despite specific searches for them.

Nobody yet knows whether or not this whole story is at all relevant to the story of the origin of biological life on our planet, but the astronomers can at least tell the biologists who work on these problems, "We can provide you with a source of the molecules you need - if they are useful to you, please use them with our compliments!"

3

The Origin of Quasars

In this chapter, we are to study the most powerful energy sources we know of in the Universe. These are located in the centres of certain classes of galaxy and we will refer to the sources collectively as **active galactic nuclei**. Before dealing with the active galaxies, we should recall the properties of galaxies in general. Normal galaxies consist of stars and gas. The dominant forms of visible matter we see in most galaxies are the stars and they determine the overall appearance of a galaxy. In the case of the elliptical galaxies, we see a spheroidal distribution of stars and the galaxy is held together by the mutual gravitational attraction of the stars for one another. In the case of the spiral galaxies, there is a central bulge, not unlike an elliptical galaxy, and a disc which is in a state of rotation. The centrifugal forces acting on the stars and gas due to the rotation of the disc are balanced by the gravitational attraction of all the stars in the galaxy. There is quite a lot of gas in the discs of the spiral galaxies but in the elliptical galaxies there is generally very little gas at all.

Until the development of radio and X-ray astronomy, most people thought that that was the end of the story - the galaxies are simply systems composed of large numbers of stars and gas clouds. Immediately after the Second World War, however, radio astronomy provided an entirely new way of looking at the Universe. Among the most spectacular discoveries of this new astronomy was the fact that a number of the most massive galaxies are extremely powerful sources of radio waves. Now, our own Galaxy emits radio waves (Figure 26) and the process by which this radiation is emitted was firmly established in the early 1950s. The emission process is known as **synchrotron radiation** and is caused by high energy electrons gyrating in the galactic magnetic field. It is a general property of charged particles that, when they are accelerated, they radiate electromagnetic radiation. In a magnetic field, electrons move in spiral paths and hence are continually accelerated towards the centres of their helical orbits. In the process of being accelerated, they emit this form of radiation at radio wavelengths. Because the electrons are of very high energy, indeed of ultra-relativistic energies, the radiation is beamed strongly in the direction of motion of the electron and this results in the characteristic properties of synchrotron radiation. Synchrotron radiation is one of the most important radiation processes in high energy astrophysics and is found ubiquitously throughout all aspects of the subject.

In Figure 26, we showed the beautiful radio map of the whole sky made by the astronomers at the Max Planck Institute for Radio Astronomy at Bonn. This radiation is entirely the synchrotron radiation of ultra-high energy electrons gyrating in the galactic magnetic field. This can be made into a very convincing story because we can measure the strength of the magnetic field in interstellar space by various astronomical techniques and we can also measure directly the fluxes of high energy electrons arriving at the Earth from nearby interstellar space by space experiments. When we put these two together, it turns out that we can explain satisfactorily the observed intensity of the galactic radio emission and many of its other properties such as its polarisation. What this means is that by making radio observations of galaxies we can determine where the high energy electrons and magnetic fields are located. These observations show that galaxies contain two other important ingredients in addition to stars and gas - **very high energy particles** and **magnetic fields**. It is apparent from Figure 26 that high energy particles and magnetic fields must be present throughout our Galaxy.

The big surprise came about when it was realised that some of the galaxies which are strong radio emitters have quite enormous radio luminosities. It turned out that galaxies, such as the massive galaxy known as Cygnus A, are about 100 million times more luminous as radio emitters than our own Galaxy. This immediately meant that there had to be enormous fluxes of high energy particles and possibly stronger magnetic fields present as compared with those in our own Galaxy. This was remarkable enough but what made matters even more complicated was the fact that the radio emission did not come from the galaxy itself

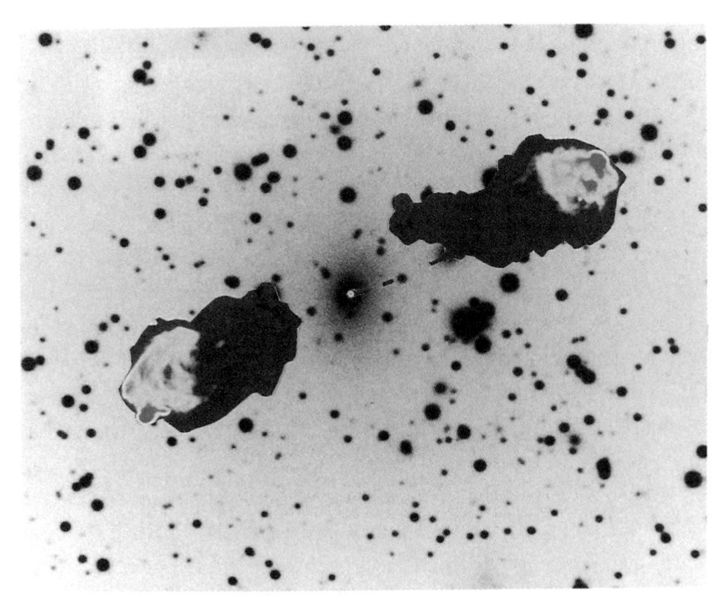

Figure 45. The radio structure of the bright radio source known as Cygnus A superimposed upon an optical picture of the same area of sky. A massive galaxy at the centre of a rich cluster of galaxies lies between the huge radio lobes and its nucleus is the energy source for the extended radio structure.

but from two enormous radio lobes located on either side of the massive galaxy (Figure 45).

It looks as if the galaxy has ejected two huge bubbles of radio emitting material from its nucleus. There have been many superb studies made of these objects using advanced radio telescopes and we now understand that the reason for these huge lobes of radio emission is that there are jets in the nucleus of the galaxy which pump high energy particles, and probably magnetic fields too, into the lobes. The superb radio picture of Cygnus A (Figure 46), which was made by the Very Large Array (VLA) in the USA, displays this process very beautifully. This means that there must be something rather remarkable going on in the nuclei of these radio galaxies - they must be able to generate enormous amounts of energy which is then expelled in the form of jets of very high energy particles. Nothing like this had been expected theoretically nor observed anywhere else in astronomy at that time.

This story was gradually beginning to unfold in the 1960s when an

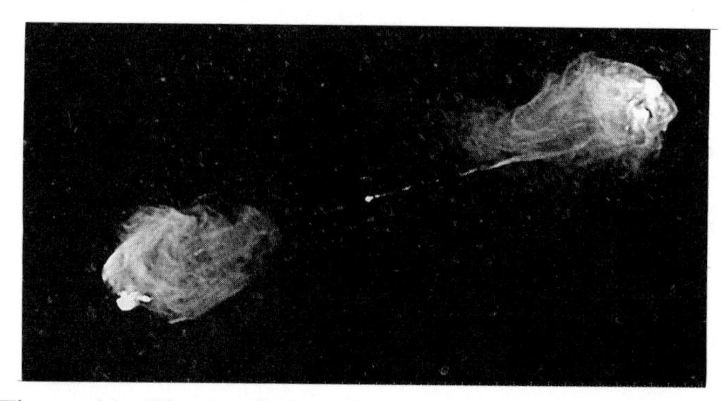

Figure 46. The detailed radio structure of the radio source Cygnus A as observed by the VLA in the USA. In addition to the radio lobes with their remarkable fine structure, there are intense "hot-spots" towards the leading edges of the lobes in which the energy density of radiation is very high. In addition, there is a compact radio source in the nucleus of the radio galaxy and there are also jets which channel high energy particles from the nucleus of the galaxy to the hot-spots.

even more remarkable discovery was made. One of the most exciting occupations of that time, in which I was involved at the beginning of my own Ph.D. research programme, was discovering the distant galaxies associated with radio sources by looking at the positions of radio sources on high quality optical photographs. This process of finding distant radio galaxies is still one of the most important tools for cosmology and indeed the most distant galaxies we know of today have been discovered by exactly this technique.

In 1960, a few radio sources were discovered by this means which appeared to be associated with stars. Their optical spectra were unintelligible. It was not until 1962 that Maarten Schmidt of the California Institute of Technology suddenly realised that the reason the spectra were unintelligible was that the whole spectrum had been significantly shifted to the red end of the optical spectrum. We will discuss what that means in more detail in the next chapter - for our present purposes, we simply note that the shift of the spectrum to longer wavelengths is a measure of the velocity of the object away from us and, in the expanding Universe, that velocity tells us how far away the galaxy is. What was amazing about this discovery was the fact that this star-like object was almost as distant as the most distant galaxies known at that time. It could not be any normal sort of star at all.

The remarkable object 3C 273 was the first of the **quasi-stellar objects**, or **quasars**, to be discovered (Figure 11). The light from the nucleus outshines the light from the galaxy by a factor of about 1000. In Figure 11, galaxies at the same distance as the quasar are the very faint smudges towards the bottom of the picture and you can see that one of these would be totally undetectable in comparison with the light from the quasar. You can also see that there is a jet pointing away from the quasar, similar to those seen powering radio sources such as Cygnus A at radio wavelengths. What made matters really exciting was the fact that the optical light of the quasars varies in intensity. In Figure 47, the variability of the radiation from two active galactic nuclei is shown, in one case the variability of the quasar 3C 345 in the optical waveband and the other the X-ray variability of the Seyfert galaxy MCG-6-30-15. You can see that in some cases, the variability can occur on timescales as short as a day or even less.

Why are these such intriguing observations? The reason is what is known technically as **causality**. Suppose we have a source which has physical size R and that it lights up very briefly indeed. What will we observe at some distance from the source? It is apparent that the light from a point on the front of the source will arrive at the observer first and that the radiation from the back of the source will arrive later by a time which is just the light travel time across the source $t = R/c$ where c is the velocity of light. Notice that the burst of radiation can last a negligible amount of time but the finite dimensions of the source mean that it will be observed as a pulse of duration roughly R/c because light travels at a finite speed. Therefore, if we observe a source to vary significantly in time t, we know that the emission must have originated from a region of physical size less than $R = ct$ or the variations would be spread out over a much longer timescale, just due to the light travel time across the source.

This is an example of what is known as **causality**. If we observe a source varying significantly in intensity in, say, one day, then we know that the size of the object cannot be any more than 1 light day in size or else the light from the back would arrive long after the light from the front.

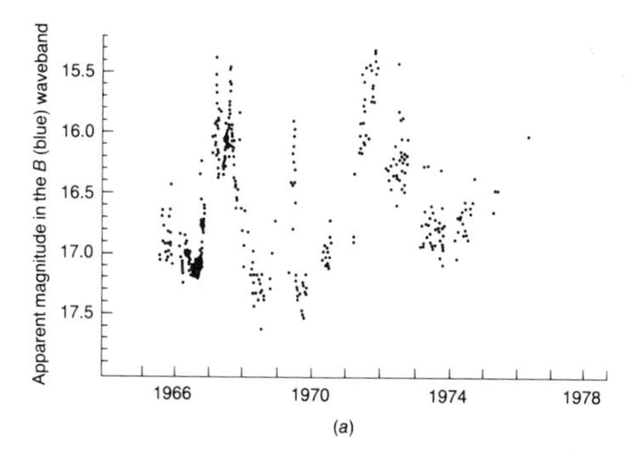

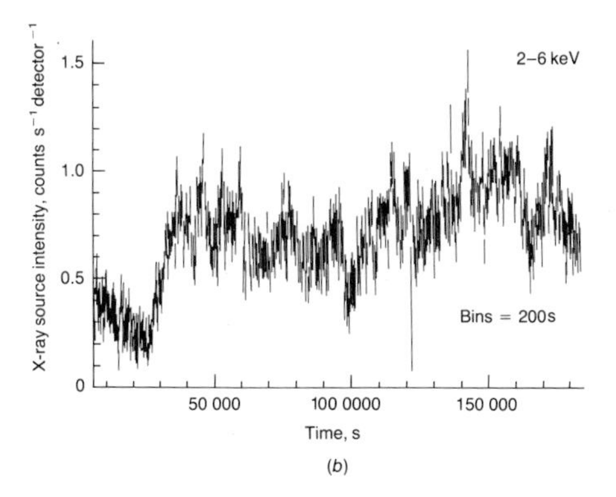

Figure 47. Examples of the variability of the emission of active galactic nuclei. (a) The optical variability of the radio quasar 3C 345 in the B waveband. (b) The X-ray variability of the Seyfert galaxy MCG-6-30-15 in the 2 to 6 keV waveband.

This means that we can obtain good limits to the sizes of the regions from which the intense emission comes. These regions are therefore very tiny indeed compared with the size of the galaxies. In fact, the enormous luminosities of quasars must originate within the very central regions of the galaxy - what we call the galactic nucleus. The quasars must therefore contain a completely new sort of energy source in which the enormous fluxes of optical, radio and X-ray emission originate within very compact regions. You have to imagine a luminosity corresponding to about 1000 times the luminosity of our own Galaxy coming from a region which is smaller than the typical distance between the stars in our Galaxy and often very much smaller. There is nothing quite like this in the vicinity of the Earth and it is somewhat reassuring that it is only in very rare classes of galaxy that the most extreme examples of this phenomenon occur. We refer to this whole class of object as **galaxies with active galactic nuclei**.

Let us now make a detour to discuss some other objects which give us important clues about how quasars might generate these huge fluxes of energy. The story turns to 1967 when Antony Hewish and Jocelyn Bell were making their first observations with a large low frequency radio telescope at Cambridge to study the twinkling (or scintillation, as it is properly called) of radio sources at long radio wavelengths. While the telescope was being commissioned, it was noted that a burst of "interference" seemed to occur at about the same time each day. After a month or so, it was found that the interference occurred at the same sidereal rather than the same solar time each day, suggesting that the interference originated from a fixed point on the sky.

The interference was observed in much more detail and the amazing discovery was made that it consisted entirely of very regularly spaced pulses with a period of about 1 second. This was the discovery of the **pulsars** and Figure 48 shows early records of two of these sources. In these early days, the sources were called LGM1 and LGM3, LGM standing for Little Green Man, because the discovery of what seemed to be some sort of celestial Morse code was totally unexpected and it was just conceivable they could have been messages from Little Green Men.

It was soon established that the only sorts of star which could produce these very sharp pulses so regularly with periods of less than a second were the **neutron stars**. Figure 49 shows a sketch of how a rotating, magnetised neutron star can produce what is observed to be pulsed radio emission. These are very, very compact stars indeed and possess very strong magnetic fields. To make a pulsar, the axis of the magnetic field cannot be aligned with the rotation axis of the neutron star. Thus, the poles of the magnet are swept around at the rotation period of the rotating star. This means that the neutron star must be rotating at a frequency of about once per second or more - a quite extraordinary speed of rotation for any star. This speed of rotation can only conceivably occur in extremely compact stars such as the neutron stars.

What are neutron stars? We can work out possible final states of stars when they die and there turn out to be remarkably few possibilities. Once a star has burned up all its nuclear fuel, it no longer has a source of thermal pressure to hold itself up. All that is left is the quantum mechanical pressure due to the fact that particles of same type are not permitted to occupy precisely the same location in space. This is a purely quantum mechanical effect and has no classical counterpart. It is exactly

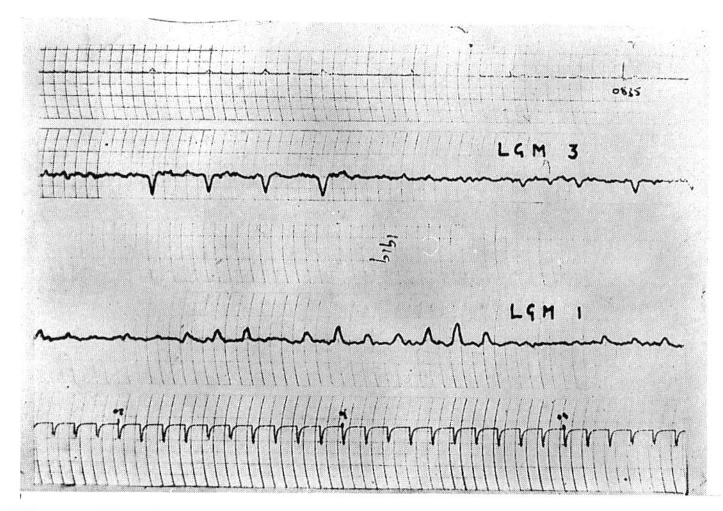

Figure 48. Early records of two of the first pulsars to be discovered. These records show the intensity of the radiation from the sources as a function of time. The time marks along the bottom of the diagram are placed at one second intervals. It can be seen from these records that the period of LGM1 is about 1 second and that of LGM3 about 3 seconds.

the same law which prevents electrons occupying exactly the same orbits within atoms which in turn leads to the whole of chemistry. Since the laws of quantum mechanics prevent particles getting too close together, this produces a pressure which is known as **degeneracy pressure**. In **white dwarf stars**, the electrons are responsible for the degeneracy pressure which enables such stars to resist collapse under gravity. These stars are quite small, having radii only about one-hundredth the radius of the Sun and they have no internal energy sources. There are large numbers of them in our own Galaxy and they form naturally as the end point of stars with mass roughly that of the Sun which evolve to the tip of the giant branch (Figure 32). These dying stars eject their envelopes non-catastrophically in the form of what are known as **planetary nebulae** and leave behind hot remnants which eventually cool to become white dwarfs.

There is a yet more compact final state, however, in which the degeneracy pressure of neutrons prevents the star collapsing under gravity. The radius of a typical neutron star is only about 10 to 15 kilometres

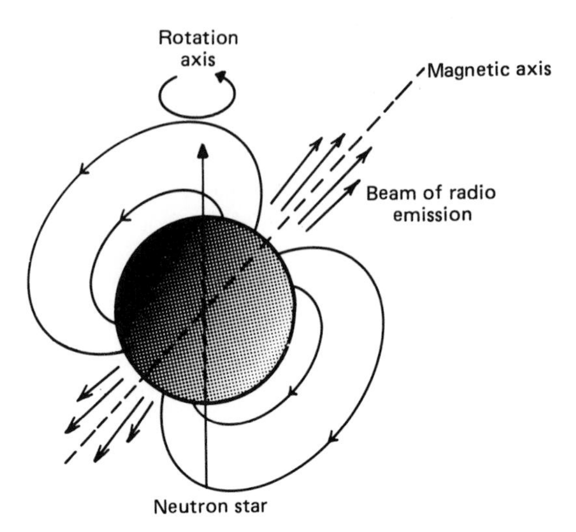

Figure 49. A schematic diagram showing how a rotating magnetised neutron star can produce pulsed radio emission. The magnetic axis of the dipole is not parallel to the rotation axis and so the poles of the magnet are swept around at the frequency of rotation of the neutron star. The radio emission is beamed out of the magnetic poles of the neutron star and consequently the distant observer observes a bright radio pulse once per orbit if the beam intersects the observer's line of sight.

and it weighs about the mass of the Sun. The density of the material inside the star is about 10^{18} kg m^{-3}. The only other place where one finds these densities is inside the nuclei of atoms themselves and so you can think of a neutron star as being a sort of giant nucleus - the whole star with all its neutrons and protons are packed tightly up against one another. That is how it is able to hold itself up against collapse. The conditions in the neutron stars are very extreme and Figure 50 provides some impression of the remarkable physical processes found in their interiors. The discovery of these stars, first suggested by Baade and Zwicky working at Mt Wilson in the USA in the 1930s soon after the discovery of the neutron, was a piece of pure luck and resulted from exploring an area of parameter space which had not previously been accessible to astronomers. For many reasons, the discovery of pulsars has opened up totally new perspectives for astronomy, far beyond their immediate

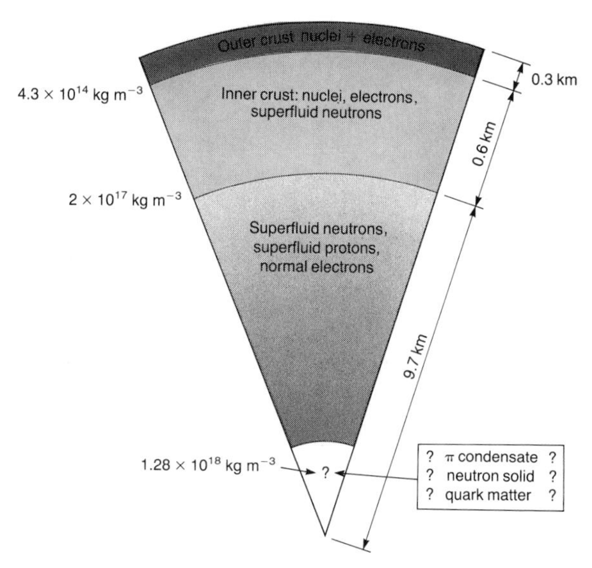

Figure 50. A representative model showing the internal structure of a neutron star of mass 1.4 times the mass of the Sun.

significance for our story here. Antony Hewish was awarded the Nobel prize for this great discovery.

Let us now turn to another great discovery made in 1971 through X-ray astronomy. X-ray astronomy was still in its infancy in 1970 when the UHURU X-ray satellite was launched. It was the first satellite dedicated to X-ray astronomy. Previous rocket experiments had shown that there exist highly variable X-ray sources. During the 1960s, X-ray astronomers were completely bemused by the fact that the sources would come and go, they would appear and disappear and it was terribly difficult to know whether the sources were real or not. The picture was clarified by one of the great discoveries of the UHURU satellite which was that, among the Galactic X-ray sources, there are objects which pulsate very rapidly at X-ray wavelengths. Figure 51 shows one of the discovery records. Again, the period of the X-ray pulsation is very short and is similar to those of the radio pulsars. This strongly suggests the presence of a neutron star in the X-ray source. The process by which the X-rays are produced, however, is completely different.

The objects would not be X-ray emitters unless there was gas present at a temperature of about 10 million K or more as can be appreciated from Figure 16. The great clue came from the fact that many of the galactic

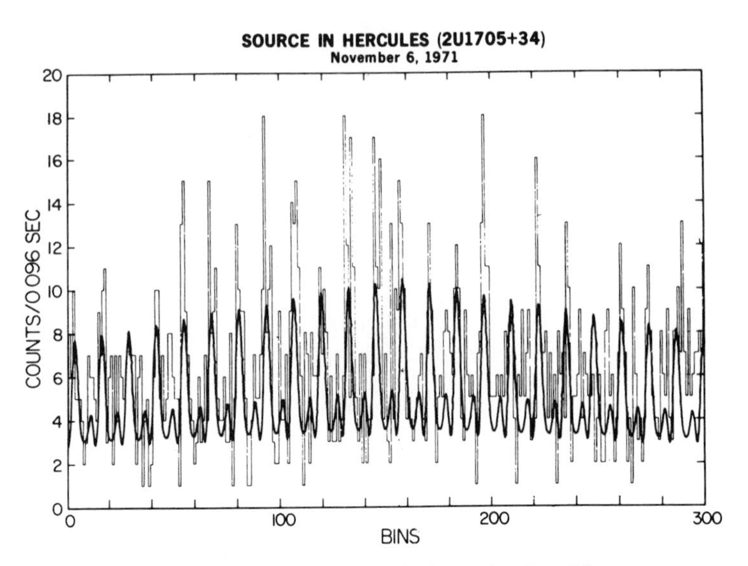

Figure 51. The discovery record of the pulsating X-ray source HerX-1 obtained by the UHURU X-ray satellite in 1971. The X-rays are emitted in regular pulses with period about 1 second, similar to the periods found among the radio pulsars.

X-ray sources turned out to be associated with binary star systems. It was soon confirmed that these are binary star systems in which one of the stars is a more or less normal star and the companion is completely invisible except as an intense source of X-rays. If the plane of the orbit of the binary lay along our line of sight, this would explain some of the highly variable activity of some of these sources since the X-ray source would be invisible when occulted by the normal primary star.

As soon as the binary nature of many of the X-ray sources was appreciated, it did not take the theoreticians long to work out that there is a very powerful means of creating hot gas in these systems. If we have a binary star system consisting of a normal star and a neutron star and we allow matter to fall from the normal star onto the very compact star, then the matter is heated up to a very high temperature indeed when it crashes onto the surface of the compact star. This process is known as **accretion** and in one way or another is responsible for the intense X-ray emission from binary X-ray sources. The matter dragged off the companion star onto the compact star is channelled by the magnetic field onto the magnetic poles of the compact star and so we can understand why there can be very hot regions close to the magnetic poles of a rotating

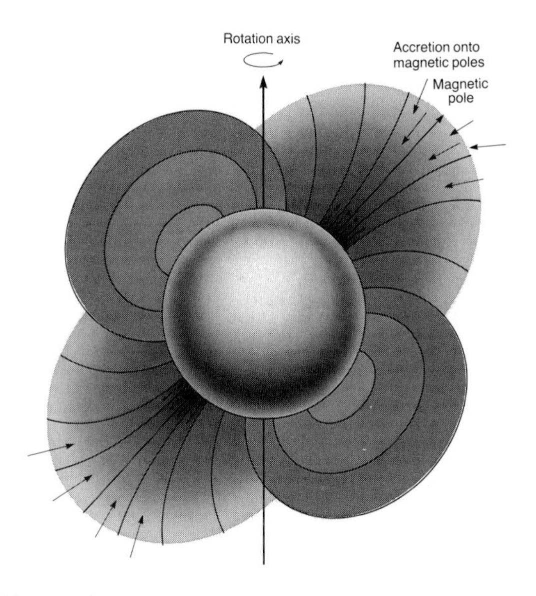

Figure 52. A schematic model illustrating how accretion onto a neutron star with a magnetic field non-aligned with the rotation axis can give rise to an accretion column which can result in pulsed X-ray emission when the source is observed at a distance. The accreting matter is channelled down the magnetic field onto the magnetic poles of the neutron star.

neutron star. As these hot spots are swept round by the rotation of the star, we observe pulses of X-ray emission at the rotation frequency of the neutron star (Figure 52). As we will show, these studies of pulsars and X-ray binaries provide a number of important clues for understanding how energy can be generated at enormous rates in active galactic nuclei.

There is, however, a major problem and that is that neutron stars are only stable if their masses are less than about twice the mass of the Sun. The same result is true for white dwarfs. Good mass measurements have now been made for eight neutron stars which are members of binary systems and they all turn out to be about one solar mass. This is very good news and is in good agreement with theory. We still have to address the question, however, of what happens to dead stars with mass greater than about twice the mass of the Sun. There is also the question of what happens if we squash a neutron star further. We recall that the neutron stars are the last stable stars we know of. If we were to squash a neutron star even further, then it would collapse, despite the enormously powerful quantum mechanical pressures which hold them up. There is no physical force that can prevent gravitational collapse to a black hole. Now a black hole is simply a hole in the structure of space-time and we will have a lot to say about them later. What I have always found remarkable is that if you take a star like the Sun and compress it so that it becomes a neutron star, then its radius would be only about 10 kilometres. The radius of a black hole of the same mass is only about 3 kilometres - in other words, the neutron stars are already almost as dense as the ultimate limiting density at which the star must collapse into a singularity. Thus, in the course of stellar evolution, objects are produced by natural processes which are very, very close to being black holes but just stop getting there because they are held up by the quantum mechanical pressure of neutrons. I have always considered that this is a good argument for the existence of black holes since not all stars can have been so accurately tuned to make neutron stars rather than black holes.

Now, what do I mean by the radius of a black hole? One very nice way of thinking about it was first noted by the Reverend John Michell in 1783 in a paper read to the Royal Society. Note that this was over 130 years before Einstein's special theory of relativity, let alone his general theory. In modern parlance, he asked "Suppose I have a star of mass M. What would the radius of that star be if the escape velocity from its surface were to be equal to the velocity of light?" You will recall the concept of **escape velocity**. It is the velocity a rocket has to have at the surface of the Earth if it is to escape from the pull of Earth's gravity. In other words, it is the velocity which enables the rocket ship to reach to an infinite distance from the Earth and to have zero velocity when it gets there. If the rocket has initial velocity less than the escape velocity, it will fall back to Earth. John Michell asks exactly the same question but now about the velocity of light. He gives precisely the correct answer which we find today from our study of black holes. There is a certain surface about the black hole at which the escape velocity is equal to the velocity of light and consequently, no light can ever escape from within that surface. In modern parlance, the radius of this surface is known as the **Schwarzschild** radius R_s of the black hole. This is such an important radius that it is worthwhile writing down the formula for it and expressing it in terms of the mass of the black hole. The Schwarzschild radius of a spherically symmetric black hole is given by

$$R_s = \frac{2GM}{c^2} = 3\left(\frac{M}{M_{\rm Sun}}\right) \quad \text{km}$$

where G is the gravitational constant, c is the velocity of light, M is the mass of the black hole and M_{Sun} is the mass of the Sun. Thus, the Schwarzschild radius of a solar mass black hole is only 3 km.

It is now time to introduce the black holes themselves - they can be thought of as coming in two families. The simplest are the **spherically symmetric black holes** or **Schwarzschild black holes** which are characterised only by their mass. All the mass is concentrated at the centre of the black hole and there is a radius about the black hole, known as the **Schwarzschild radius**, which, as we have shown, acts as the surface of the black hole. It is the point of absolutely no return for radiation as we described above. The reason the object is called a hole is that if matter or radiation get too close to it, they inevitably spiral into the hole despite the fact that they may have some rotation about the black hole. According to the general theory of relativity, when the matter approaches too closely to the Schwarzschild radius, rotation is no longer capable of preventing the matter from collapsing to the centre and, in fact, positively helps the matter to fall in. From the astrophysical point of view, the important point is that there is a last stable circular orbit about a Schwarzschild black hole which lies at a radius which is three times the Schwarzschild radius. Matter can take up stable circular orbits with any radii greater than this but as soon as the orbit has radius $3R_s$ or less, the matter is dragged inexorably into the hole (Figure 53 (a)).

The other form of black hole is the **rotating or Kerr black hole**. In general, we would expect black holes to have some rotation (or to put it more properly, angular momentum) because the matter out of which they form is likely to have some net rotation, for exactly the same reasons that the protostars are likely to rotate (see Chapter 2). In addition, we expect that they will pick up angular momentum because the infalling matter may well have angular momentum and this speeds up the black hole. There is, however, a maximum amount of angular momentum which a rotating black hole can possess - if it had any more, it would spin so rapidly that it would never have formed in the first place. It would have been torn apart by centrifugal forces. For those interested in the technicalities, the upper limit to the angular momentum J is given by the formula $J = GM^2/c$.

The rotating black holes are even more exotic than the spherically symmetric black holes. There is again a "surface of infinite redshift" from inside which radiation and matter cannot escape. In the case of the maximally rotating black hole, this radius is much smaller than in the non-rotating case and is equal to half of the Schwarzschild radius. Outside this surface, there is a volume in the form of an ellipsoidal shell

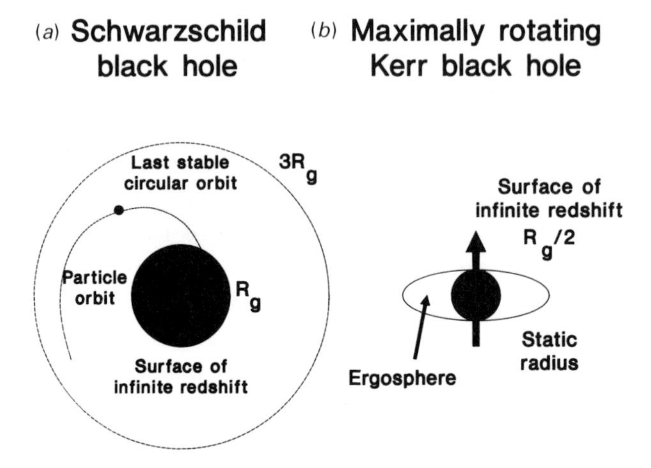

Figure 53. Illustrating the structures of (a) spherically symmetric (Schwarzschild) black holes and (b) rotating (Kerr) black holes. In the latter case, the black hole is rotating as fast as possible.

known as the **ergosphere** in which particles must corotate with the black hole (Figure 53(b)). From the point of view of energy generation, the important thing is that the last stable orbits about the black hole come very much closer into the black hole and therefore the amount of energy that can be extracted is greater than in the case of the non-rotating black hole. We will describe the results of these calculations in more detail in a moment.

Now, this is all very interesting but how does it have anything to do with quasars and active galactic nuclei? At first glance, it looks as though we have produced objects which are very good at getting rid of matter and energy rather than emitting it! The clue is again provided by the binary X-ray sources. As we demonstrated above, the source of energy in these systems is the accretion of matter from the normal primary star onto the compact secondary star. It turns out that this is an extremely efficient means of producing energy, provided the star is compact enough. All we do is to take matter and drop it from a very great height onto the surface of a neutron star. If I do that experiment as a lecture demonstration, it is not all that impressive. If I drop a cannon-ball onto the floor, it makes a great clatter. What happens is that the ball loses the kinetic energy it gains in falling when it collides with the floor and some of that energy has been converted into the noise we hear, some has gone into bending the floor and heating it up but it is not really terribly impressive. When we are dealing with collapse of matter onto a neutron star, however, the matter acquires a quite enormous velocity before it hits the surface and sticks there. In fact, a very simple sum shows that the matter would pick up a velocity which is a significant fraction of the velocity of light. If you do this sum carefully, you will find that if the matter just collapses onto the surface of a neutron star, as much as about 5% of the rest mass energy of the infalling matter can be converted into heat. This is a quite enormous amount of energy. We can compare it with the best that we can obtain from nuclear fusion reactions. When we described the nuclear reactions which power the Sun, you will recall that we were able to extract a maximum of just less than 1% of the rest mass energy. Here we are doing at least five times better.

In the above example, the matter collapsed onto the surface of a neutron star. What happens if the matter collapses into a black hole? How much energy can we, in principle, extract before it disappears inside the Schwarzschild radius? Let me give the figures, first of all, before we ask how the black holes might actually be able to do something useful with the energy. The spherically symmetric black holes enable us to recover about 6% of the rest mass energy of the matter falling into the black hole, i.e. an efficiency about 10 times greater than nuclear energy. If the black hole is allowed to rotate as fast as it possibly can, about 42% of the rest mass energy of the infalling matter can be extracted for useful purposes. This is a quite fantastic amount of energy. It is roughly 50 times more than can be produced by nuclear energy. Finally, we could ask how much of the rotational energy of a maximally rotating black hole could be made available if we tried to tap it. The theoretical answer is about 29% of its rest mass energy. Clearly, we have access to very powerful energy sources in rotating black holes.

Notice, however, that we have obtained much more than simply energy. We have found a source of energy which can liberate energy on the shortest possible timescale for an object of a given mass. This is because most of the energy is generated from roughly the last stable orbit about the black hole and that means that we can expect that there could be variations in the intensity on timescales of the order of the light travel time across the black hole. Since the Schwarzschild radius R_s is the smallest meaningful physical scale which we can associate with the mass M, R_s/c must be the shortest timescale on which energy is produced.

So we obtain two things simultaneously from the black holes. We get enormously powerful sources of energy in the sense that one can almost liberate the rest mass energy of the matter falling into the black hole and also we get the shortest possible timescale for an object of that mass. It is all beginning to hang together and the question is - can we apply these ideas to the most energetic phenomena we know of in the Universe?

Now, you would have every right to ask the question, "Suppose, instead of falling onto a solid surface, such as the surface of a neutron star, the matter falls into the black hole which has no solid surface - how are we able to release all this energy in practice?" At first sight you might say, "Well, if the matter falls into a black hole, then that's not much good - we have simply lost all the matter and not extracted any energy at all!" This is another area in which observations of X-ray binaries have provided the essential clue because it is very unlikely that the matter can ever fall directly into the black hole. It is the old problem we encountered when we tried to make stars - the matter must possess some residual rotation about the black hole if it falls in from a great distance and, just as in the case we discussed in Chapter 2, we have to find some way of getting rid of the rotation of matter with respect to the black hole or the neutron star if it is to reach the "surface" of the object. Now, there is a very convincing way in which we know that this can happen and that is if the matter forms a disc about the black hole. This will happen very naturally because matter will tend to collapse along the rotation axis of the infalling material. Matter will only be able to move into a closer orbit about the black hole by losing energy and it can do this by friction and viscosity in the accretion disc. This is a beautiful piece of astrophysics in which the gas loses energy by friction in the disc and this achieves two highly desirable things - first, energy is dissipated and this heats up the disc and, second, the matter gradually drifts in towards the black hole (Figure 54). It is by this means that the infalling matter can deliver up a large fraction of its rest mass energy as heat.

Before we go back to the real world of X-ray sources and active galactic nuclei, we need one more idea. If the luminosity of any source is too great, the radiation pressure of this luminosity will blow away the matter which is fuelling the neutron star or black hole. This is simply the pressure associated with the enormous energy flux coming from the nucleus (Figure 55). This is a very famous calculation first performed by Arthur Eddington in which he showed that there is a maximum luminosity for any source before it blows the infalling matter away. What is of special interest is that this limiting luminosity only depends

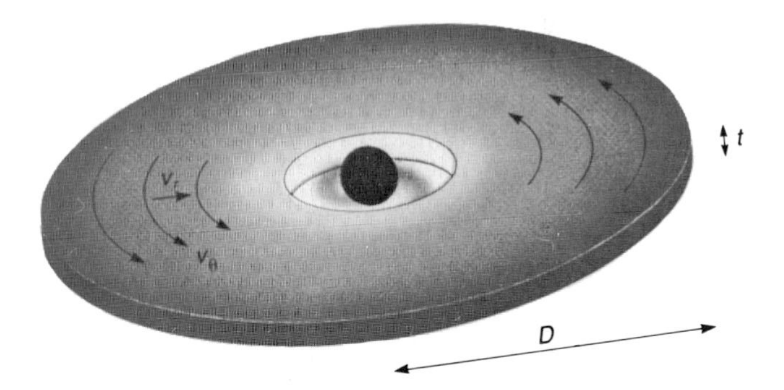

Figure 54. Illustrating the structure of a thin accretion disc about a black hole. The material drifts in towards the black hole at velocity v_r. The dissipation of energy associated with the outward transfer of angular momentum in the disc increases towards the central regions. In the case of a black hole, the last stable orbit about the black hole occurs at $r = 3R_s$ where R_s is the Schwarzschild radius.

upon the mass of the object. For those who are interested in checking these sums for themselves, the Eddington limiting luminosity L_{Edd} is

$$L_{Edd} = 10^{31}\left(\frac{M}{M_{Sun}}\right) \quad \text{watts}$$

One of the rather remarkable things about the Galactic X-ray sources is that their X-ray luminosities extend up to about the Eddington limiting luminosity and more or less stop there, assuming that the masses of the accreting stars lie between about 1 and 10 times the mass of the Sun. This is very good evidence that, in the case of the X-ray sources at least, there exist objects which can radiate more or less up to the theoretical maximum luminosity without too much difficulty.

Have astronomers discovered any black holes? That is a very intriguing question. There are now at least three rather good candidates discovered by X-ray astronomers in which there are known to be invisible companions present in binary X-ray sources. These invisible companions seem to be rather massive - meaning about 10 times the mass of the Sun. The mass estimates of these unseen companions all exceed the upper limit for white dwarfs and neutron stars of about twice the mass of the Sun. Therefore, if these objects really are as massive as 10 times the mass of

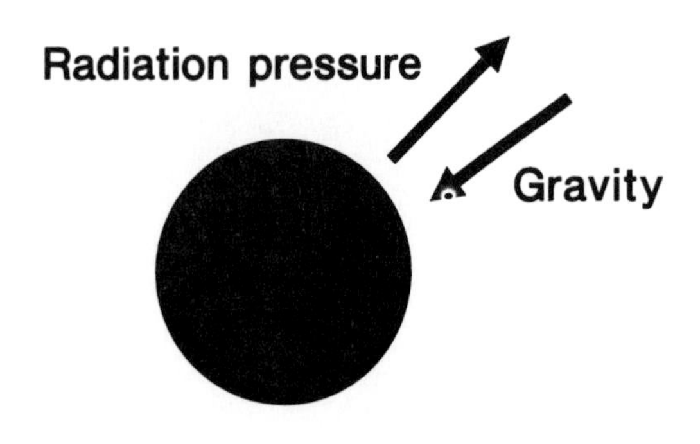

Maximum luminosity ▪ 10^{31} M/M(Sun) W

Figure 55. Illustrating the origin of the Eddington limiting luminosity.

the Sun, they can only be black holes. This a very important and exciting area of astrophysics and we need to confirm very urgently that indeed there are black holes in these binary systems.

The other area in which the case for the presence of black holes can be made rather convincing is the case of the active galactic nuclei. Let us now try to put together all the ideas developed in the preceding paragraphs to understand the phenomena we observe. Let us take one beautiful example - the case of the active galaxy NGC4151 which we described in Chapter 1 (Figures 9 and 10). A number of my European colleagues carried out a remarkable set of observations using the ultraviolet space telescope known as the International Ultraviolet Explorer (IUE). This galaxy contains one of the most active nearby nuclei and there are rather dense gas clouds in the vicinity of the nucleus. What they discovered was that these gas clouds are heated and excited by the ultraviolet radiation emitted very close to the nucleus itself. The exciting feature of the set of observations was that they were able to measure the time delay between a burst of radiation occurring in the nucleus and that burst stimulating the clouds surrounding it to radiate. This is excellent news because it means that we can estimate the distance of the clouds from the nucleus since the radiation propagates at the velocity of light. This is a wonderful observation because we can now work out how much mass there must be within the orbits of these gas clouds. It is exactly the same calculation which enables us to measure the mass of the Sun if we know the Earth's distance from the Sun and its orbital velocity. All we have to do is to balance the centrifugal force acting on the clouds by the gravitational attraction of the nucleus

$$\frac{mv^2}{R} = \frac{GMm}{R^2}$$

and hence one finds directly the mass of the nucleus

$$M = \frac{v^2 R}{G}.$$

What emerged was that, in the nucleus of the galaxy NGC4151, there must be a compact object with mass at least 100 million times the mass of the Sun. Similar calculations have now been made for a number of other active galactic nuclei - the most luminous emitters have masses which lie in the same range, 1-1000 million times the mass of the Sun.

We have derived three important relations concerning the properties of black holes as energy sources in active galactic nuclei which provide strong constraints upon models. These are:

1. Accretion provides very efficient energy production with efficiencies amounting to 5 to 40% of the rest mass energy of the infalling material.
2. The shortest timescales of variability for an object of mass M is given roughly by $R_s/c = 10^{-5}(M/M_{\text{Sun}})$ seconds.
3. The maximum luminosity of the source is limited by radiation pressure according to the Eddington limit $L < 10^{31}(M/M_{\text{Sun}})$ watts.

How well do known active galactic nuclei satisfy these criteria? The answer is displayed in Figure 56 which shows mass estimates for a range of active galactic nuclei derived from both dynamical arguments and from the variability of the emission from the nuclei. It can be seen that these mass estimates are in reasonable agreement. Then, using these masses, each active nucleus can be plotted on a mass-luminosity diagram to find out whether or not it approaches the region which is forbidden according to the Eddington criterion. It can be seen from Figure 56(b) that the active galactic nuclei are close to but do not fall in the forbidden region. These diagrams indicate that even the most extreme quasars can be accommodated within the simplest of black hole models. My own view is that these arguments are wholly convincing that there must be supermassive black holes in active galactic nuclei.

Can we now put all of this together to explain the many different

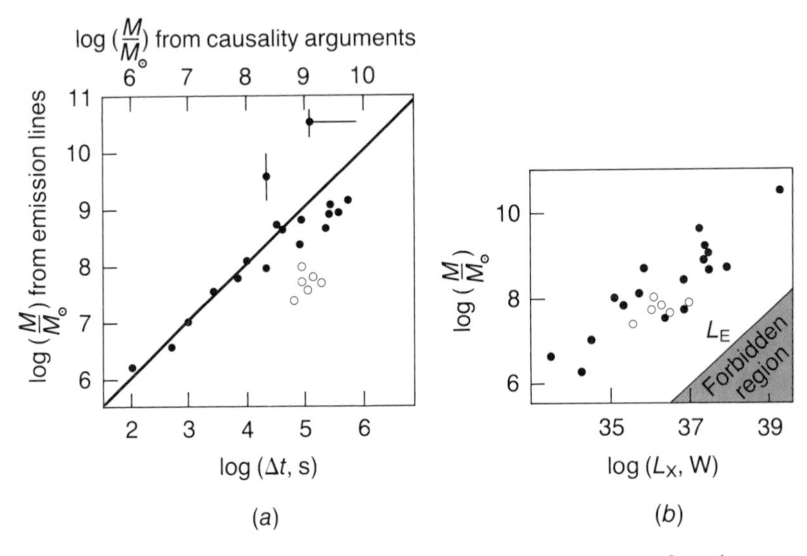

Figure 56. (a) Comparison of the mass estimates of active galactic nuclei from the variability of their X-ray emission and from dynamical estimates. The solid dots are quasars and type 1 Seyfert galaxies. The open circles are type 2 Seyfert galaxies. (b) Comparison of the inferred masses and luminosities with the Eddington limiting luminosity. It can be seen that all the points lie well away from the Eddington limit. $M_{\odot}$ is the mass of the Sun.

phenomena we see in active galaxies? The honest answer is "Not really!" - the general arguments are strong but the more detailed arguments are difficult and a great deal of guesswork is needed in order to explain everything we see. For example, we know that there are dense gas clouds in orbit about the nucleus but where does the gas come from? We are not certain. We know that there are radio jets coming out of the nucleus of the galaxy but how do they manage to do this? We can guess that it will be simplest for the radiation to escape along the axis of a rotating black hole but we do not yet have a really convincing physical model for the accretion discs about the most active nuclei or for the environments of the black holes. We do not know how black holes are able to accelerate particles to very high energies. We know that they are certainly able to do this but the physical mechanism involved is not really known. Not only do we have to produce the particles but we have to liberate them in beams or jets at very high speeds. One of the most amazing results

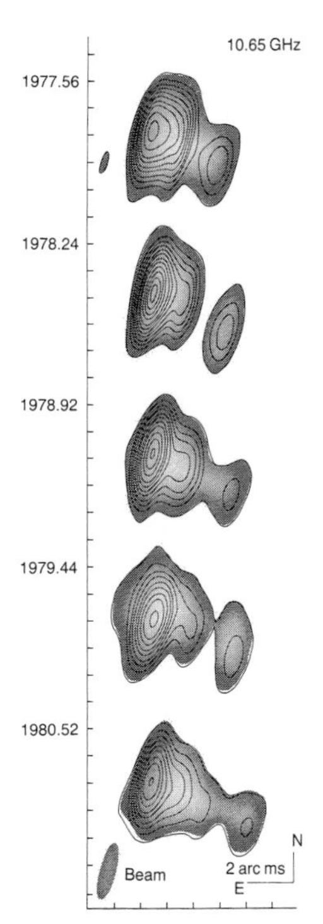

Figure 57. Very Long Baseline Interferometry images of the nucleus of the radio quasar 3C 273 for the period 1977 to 1980. The radio component moves a distance of about 25 light years in a period of three years implying a superluminal expansion velocity of about eight times the speed of light.

of modern astrophysics has been the discovery that some of these jets seem to be escaping from the nucleus at velocities which are greater than the velocity of light (Figure 57). Now, according to Einstein's special theory of relativity, this cannot happen. Almost certainly what we are seeing is some sort of aberration effect whereby the jet is moving at a very high but subluminal velocity and it simply appears to be moving sideways more rapidly than the velocity of light because the source of the radiation is rushing towards us. In Figure 58, I have put together a sketch of what the nuclear regions of a quasar may look like but this is highly schematic and there is certainly no guarantee that the structure I have sketched is stable.

So, where have we got to? What is the origin of quasars? I think you will find that most astronomers will agree that there must be supermassive black holes in active galactic nuclei in order to explain their extreme

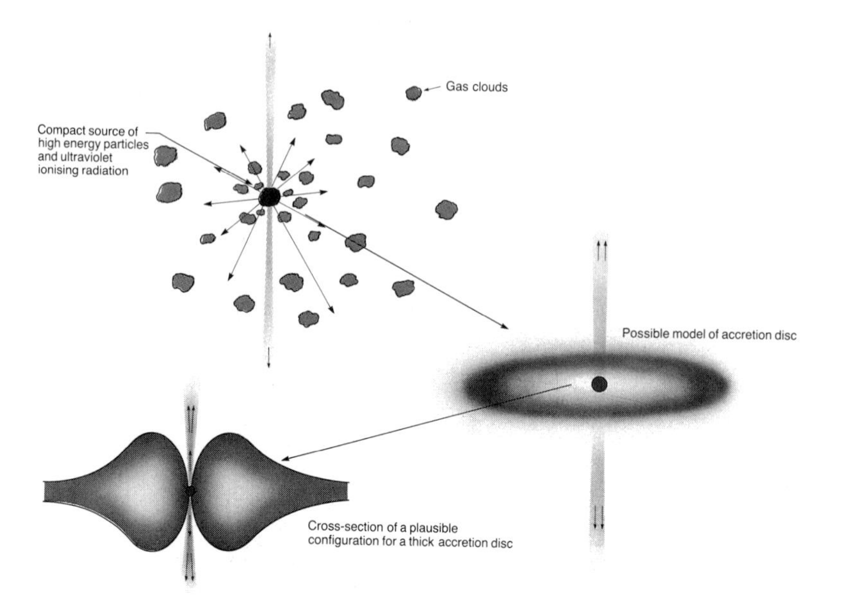

Figure 58. A schematic model showing the necessary ingredients of a model of an active galactic nucleus. There must be a compact source of radiation and high energy particles in the nucleus. The nucleus is surrounded by gas clouds which are heated and excited by the ultraviolet radiation from the nucleus. There must be some mechanism responsible for producing the beams of particles. The inserts show a possible structure for the accretion disc and the thick disc surrounding the black hole. These may be responsible for producing the narrow beams of high energy particles. It is not at all certain that this configuration of a thick disc is stable.

properties - specifically, the black holes have all the properties that we need to explain the luminosities and timescales of these phenomena. But how we convert these general rules into the definitive theory of quasars is probably for the next generation of scientists. The implications of these studies are very far-reaching because we have discovered very compact supermassive objects and in them the effects of gravity are no longer weak but are very strong indeed. We have the opportunity in these systems to study the behaviour of matter in strong gravitational fields that we cannot find anywhere else in the Universe. One of our great goals is to understand more about the behaviour of the laws of physics under such extreme physical conditions. This is where we need your help - you have to help us to solve the great mystery of how the most powerful energy sources in the Universe are able to produce the amazing range of new phenomena which are baffling the present generation of astronomers.

4

The Origin of Galaxies

In Chapter 1 we described how the large scale structure of the Universe is defined by the distribution of galaxies. We thought of the galaxies as being simply points which trace out that large scale distribution. What we want to study in this lecture is how to make galaxies and this problem turns out to be inextricably mixed up with the origin of the larger scale structures. In Chapter 2, we described the life cycle of stars - their birth in dense dusty regions of interstellar space, their long lifetimes as ordinary stars and their violent deaths - what I have called the great cosmic cycle of birth, life and death of stars. So, we already know quite a lot about the internal workings of the constituents of galaxies. There is a continual interchange of matter between the stars and the interstellar gas. The stars form in dense regions of interstellar gas and, when they die, they return processed material to replenish the interstellar medium. As a result, there is a great deal of rich structure in the interstellar medium and an enormous range of phases of the gas, ranging from very cool gas in the dusty regions to very hot gas which has been heated up by the explosions of stars. As a result, once a galaxy gets going, we can build models of the evolution of its stellar content once we know how the rate at which stars form depends upon local physical conditions. I am conscious that I am brushing under the carpet a great deal of uncertainty and complexity but, in principle, we can build models of galaxies which bear some resemblance to the objects we observe.

The big problem is, however, "How did it all get started?" In other words, how was the gas out of which the galaxies formed brought together in the first place so that the processes of star formation and the great cosmic cycle can get underway? That is the problem we study in this chapter. We will assume that, once the stars form, the galaxy will look after itself - that's what we have talked about in the first three chapters. Our goal does not, therefore, appear to be too ambitious at first sight - all we have to ensure is that there are significant perturbations or enhancements in the background density in the Universe out of which large self-gravitating systems such as galaxies and clusters of galaxies can form. What this means is that we need to find physical mechanisms for producing density enhancements out of the diffuse matter in the Universe. It sounds terribly easy but in fact it results in some very profound problems.

To appreciate how these problems come about we have to understand the background against which we are trying to clump the matter together. In other words, we need to understand the large scale motions that are present in the Universe. This is one of the most remarkable stories of 20th century science and it began in the early years of this century. At that time, it was not at all clear whether the objects we now know as the galaxies belonged to our own Galaxy or whether they were distant "island universes". Astronomers and philosophers of the 19th century had speculated that the galaxies were distant island universes similar to our own Galaxy but there was no direct proof of that. The proof of the extragalactic nature of the galaxies came following a remarkable debate which ended in about 1920 when the American scientist Edwin Hubble established to everyone's satisfaction that the galaxies are indeed distant systems similar to our own Galaxy. This was the beginning of extragalactic research and observational cosmology - most of Hubble's subsequent career was devoted to studying the large scale distribution of galaxies in the Universe. In 1929, he discovered the famous law which bears his name - **Hubble's Law**. He found that the distribution of galaxies is not stationary but that the further we look out into the Universe, the greater the velocities of the galaxies away from our own Galaxy. In fact, there is a linear proportionality between the distance of the galaxy and the speed at which it is rushing away from us - this is Hubble's Law. Figure 59 is a modern version of that relation. Along the horizontal axis is plotted the apparent magnitude of the galaxies which is just a logarithmic measurement of the observed intensity of the galaxy and which is proportional to the logarithm of the distance of the galaxy, provided all the galaxies plotted have the same intrinsic luminosity. Up the vertical axis is plotted the recessional velocities of the

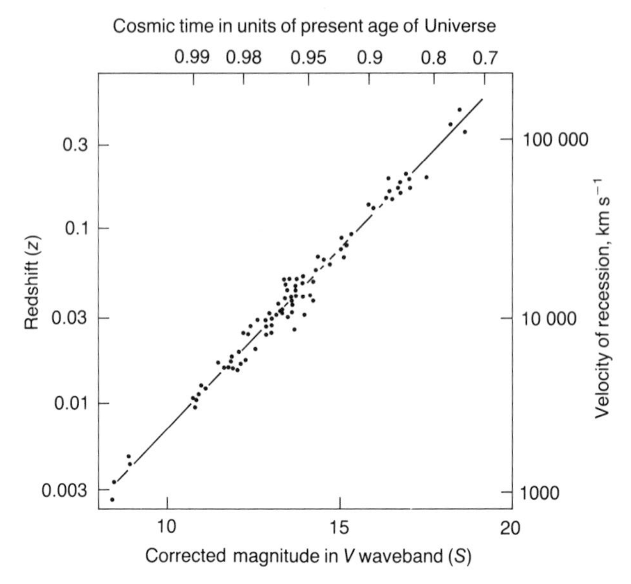

Figure 59. A modern version of Hubble's relation for the brightest galaxies in clusters of galaxies. In this logarithmic plot, the corrected apparent magnitude V is plotted against redshift, z. The apparent magnitude is defined as $V = \text{constant} - 2.5 \log S$ where S is the flux density in the V waveband. The straight line is what would be expected if S was proportional to z^{-2}. The small dispersion about this line shows that the brightest galaxies in clusters must have remarkably standard properties and that the distances of the galaxies are proportional to their recession velocities or redshifts.

galaxies, i.e. their velocities away from us. You can see that the points lie beautifully along a straight line at 45 degrees showing that velocity is exactly proportional to distance. I can assure you that correlations never come any better than this in cosmology.

Before we look into what Hubble's Law means let us look at one other piece of information and that is the question of whether or not our Universe looks the same in all directions on the very largest angular scales. Now, you will remember that, in the first chapter, we agonized a great deal about the fact that the distribution of galaxies is very irregular - there are large holes or voids as well as sheets and filaments made up of galaxies. The question is, if we take large enough regions of the Universe, does the distribution eventually end up being smooth or is there still irregularity on the very largest scales? If we take averages over large enough volumes so that lots of voids and clumps are included in the average, then it does appear that the Universe is reasonably uniform in different directions. You can appreciate this by looking at the unobscured regions of Figure 13 which show the distribution of galaxies in the Northern Galactic Hemisphere. Although the distribution is irregular, one bit looks very much like another. If we think in terms of our sponge model of the Universe, what we mean is that one bit of the sponge is very much like any other bit of the sponge, provided we take averages over large enough numbers of holes.

Nowadays, there is more evidence about the uniformity of the large scale structure of the Universe. We can, for example, select only very distant objects by studying large samples of the radio sources which we described in Chapter 3. Figure 60 shows the positions of the brightest 3000 radio sources in the northern sky. I have again bent the coordinates so that equal areas on the sky correspond to equal areas on this picture. You can see a rather prominent hole in the middle of the distribution but that is there because that bit of sky was not surveyed in the radio survey. If we exclude that region from our studies, we can show that the distribution of the points is entirely consistent with the radio sources being placed randomly on the sky. Any apparent "constellations" or clusters are simply due to the random superposition of points. This strongly suggests that the distribution of distant objects is the same in all directions on the sky.

We have, however, even better information about this key question from the distribution of the microwave background radiation on the sky. We are going to have a great deal to say about this radiation in the last chapter but, in a word, it is the cool remnant which we can detect today of the hot early phases of the Universe. We have already shown the very beautiful radio maps of the sky obtained by the COBE satellite in Figure 24. That picture is dominated by the slight increase and decrease in the intensity of the background radiation due to the Earth's motion through the microwave background radiation. The COBE scientists have taken out the effects due to the motion of our Earth through this radiation with the results shown in Figure 61. The plane of our own Galaxy can be seen clearly now. Except in the direction of the Galactic Plane, however, the intensity of the microwave background radiation is the same in all directions to better than one part in 30 000. Now, you may say, "Well, what on Earth has the microwave background radiation to do with the Universe of galaxies?" - that is one of the key ideas we have to tackle in

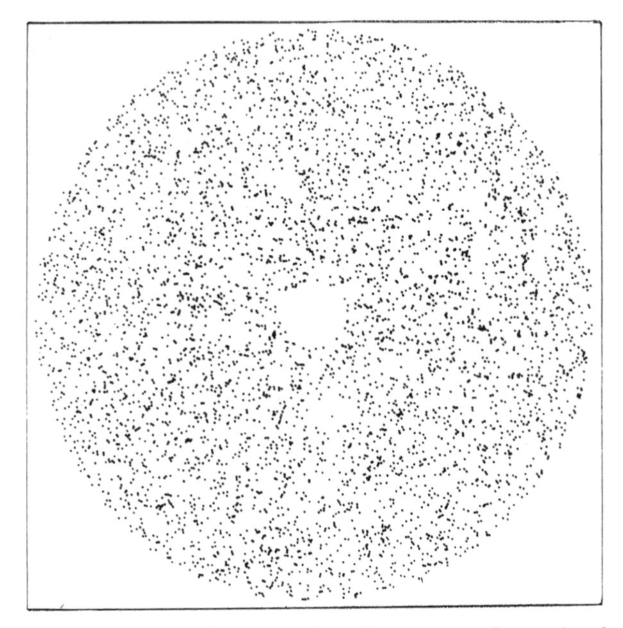

Figure 60. The distribution of radio sources from the fourth Cambridge catalogue in the northern hemisphere. In this equal area projection, the north celestial pole is in the centre of the diagram and the celestial equator around the circumference of the distribution.

this chapter. For the moment, I ask you to accept that this is indeed the background against which we have to make the galaxies. The fact that the distribution is so uniform in all directions tells us that it is a very good approximation to begin with models of the Universe which are the same in all directions. Technically, we say that on the large scale the Universe is isotropic.

Let us now return to Hubble's Law. When we take together Hubble's Law and the isotropy of the Universe, we find that Hubble's Law has a much deeper meaning. In Figure 62, I show a uniform distribution of galaxies expanding uniformly. Now, the trick is to concentrate your attention upon any one of the galaxies and ask how the other galaxies have to move relative to your chosen galaxy so that the distribution of galaxies remains uniform. It is simplest to concentrate upon a line of galaxies. You will notice that, in the uniform expansion, the further away a galaxy is from your chosen galaxy, the further it has to move in a given interval of time in order to preserve uniformity. In fact,

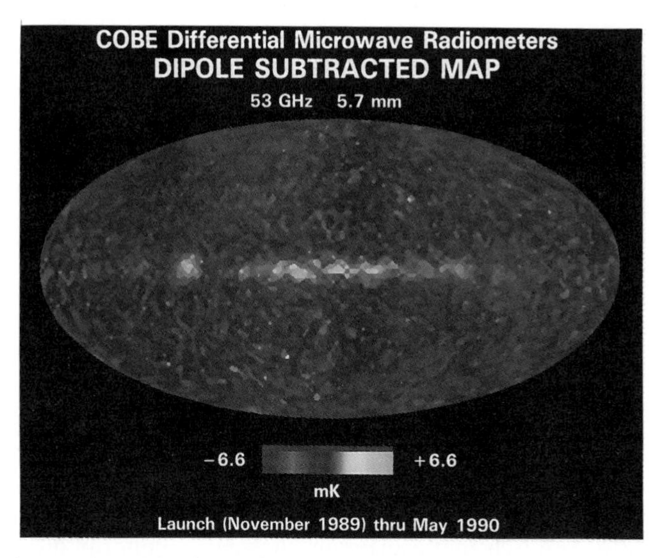

Figure 61. The intensity distribution of radiation over the celestial sphere at a wavelength of 5.7 mm as observed by the COBE satellite once the effect of the Earth's motion through the microwave background radiation has been removed. Away from the Galactic Plane, there is no evidence for any anisotropy (unevenness) in the distribution of the radiation.

by inspection of Figure 62, you can see that in a uniform expansion, the velocity of recession of a galaxy has to be exactly proportional to distance, i.e. exactly Hubble's Law. It is a rigorous statement that, in a uniformly expanding Universe, the further away a galaxy is then the faster it has to be moving away from us in order to guarantee that the Universe remains isotropic and uniform. The other remarkable thing is that, in this model, all observers see an expanding Universe and they all see the same Hubble's Law if they make their observations at the same time. This is the real significance of Hubble's Law - it tells us that the Universe as a whole is expanding uniformly. Notice that everyone thinks that they are at the centre of the Universe and when we contract the Universe back in time, everyone was there at the beginning.

One of the immediate consequences is that the Universe must have started off very compact and very hot because, as we saw in the explosion of a supernova, matter cools as it expands. This means that the early phases of the Universe must have been very hot indeed and that is why we call this model the **Hot Big Bang** model of the Universe. We are

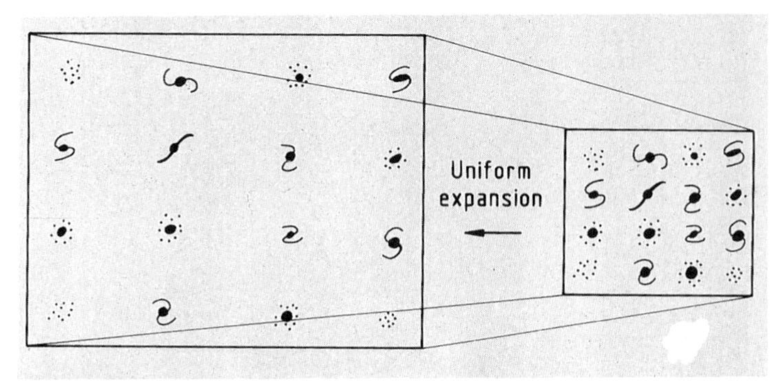

Figure 62. A schematic diagram showing a uniform distribution of galaxies partaking in a uniform expansion of their distribution.

going to look into these hot early phases in much more detail in the last chapter. In this chapter, we restrict attention to the problem of forming galaxies in the expanding Universe. We will forget all about the extremely hot early phases and consider only the relatively recent Universe.

Now, it looks as if the programme is pretty simple from this point on. We have established that the Universe is expanding and that it is rather uniform. What the theoretician now does is to say, "Well, suppose I start with a completely uniform, expanding Universe. Let me now put in small imperfections into my uniform expanding Universe and see whether these perturbations grow or not as the Universe expands." We start off being optimistic because you will remember that in the second chapter we showed that gravity acting on a stationary cloud on a large enough scale is a very good way of producing exponential growth of the density of that region. This is almost certainly the way in which stars form from the diffuse interstellar gas. We might hope that the same thing would happen in the Universe in order to start the formation of galaxies. This is where the big problem arises - it does not.

The reason why these little perturbations do not grow exponentially is very interesting. The little fluctuations do grow but only very slowly. This is because, as soon as the density of the little region tries to increase under the influence of its own gravity, most of the good work is undone because the Universe as a whole is expanding and decreasing the density of the region as a whole. As a result, the perturbation feels a decreasing gravitational pull and the density increases very slowly. We can illustrate this problem by carving out a little region of the Universe and squashing it to slightly higher density than the background density. We can then work out how the density of that region grows relative to the background expanding Universe by treating the little region as a closed Universe of slightly higher density than the background. It turns out that this model provides an exact description of the growth of the density contrast of the region which we describe below.

Let us show what the exact result is. Suppose we have a set of points which expand with the Universe. Then, we can define the expansion of the Universe in terms of the relative separation of these points. If we suppose they are separated by 1 unit at the present time, then they were separated by some smaller number of units R in the past. This number R is very useful indeed in cosmology and is known as the **scale factor** of the Universe. How it changes with time defines the expansion of the Universe. If the Universe is sufficiently dense, it turns out that the little density enhancements grow as

$$\frac{\delta\rho}{\rho} \propto R$$

where $\delta\rho$ is the density increase over the average density ρ. I hope you don't mind this little bit of mathematics because it is such a beautiful result and it has very profound implications as we will see in a moment.

Now why is this such a problem? What we might have hoped is that the galaxies would form from random perturbations in the early Universe. These must occur statistically just as in our illustration of laying down radio sources at random on the sky but this will only work if these fluctuations are able to grow exponentially so that they attain a finite density in a finite time. This is the nice thing about exponential growth. Linear growth, of the type described by the above formula, is just too slow to make finite size perturbations from infinitesimal perturbations in the early Universe. The theoreticians have tried very hard to find a way round this problem but no solution has been found.

There is, however, an even bigger problem and that is the effect these growing perturbations would have upon the microwave background radiation. If we take our Universe backwards in time, it gets hotter and hotter and if we squash it until the scale factor, i.e. the distance between galaxies, is just over one-thousandth of what it is now, then the temperature of all the diffuse matter and radiation in the Universe increases to about 4000 K. We recall that hydrogen is by far the most abundant chemical species in the Universe. At this temperature, all the hydrogen in the Universe is broken up into its constituent protons and electrons by the hot tail of high energy radiation in the background radiation which is now glowing like a star just a bit cooler than the Sun. This breaking up of hydrogen into its constituent particles is known as **ionisation** and this has important consequences for the interaction between radiation and matter in the Universe. The most important consequence is that matter in the form of free electrons and protons is what is known as a plasma and it has quite different electrical properties from neutral cold matter. In particular, the coupling between the radiation and the matter is now very strong indeed. The negatively charged electrons scatter the background radiation very effectively and therefore we can think of the matter and radiation as now being tightly tied together or coupled by these interactions.

You can now understand what is going to happen. We can work out how big the fractional perturbation in density has to be in the past in order to ensure that we make galaxies by the present time. As an example, say we have to make galaxies by now. We know that the very least we have to achieve is that the density contrast by now is about 1. Now, if we squash the distance between galaxies by a factor of 1000, we know that these perturbations must have been at least one part in a thousand of the background density because of the law of growth of the perturbations which we have just mentioned. This is where the ionisation of the Universe at that time becomes important because the matter was then coupled to the radiation and therefore, if there are fluctuations in the matter distribution, we also expect to observe similar perturbations in the radiation as well. Detailed calculations show that we would expect to see fluctuations in the radiation on the sky at least at the level of about one part in a thousand and this is in clear contradiction with the extreme isotropy of the microwave background radiation.

This is perhaps the most difficult problem for the simplest of pictures of how galaxies might have formed in the expanding Universe. If there is only ordinary matter present in the Universe, it is very difficult to avoid the conclusion that we ought already to have seen traces of the formation of galaxies and larger scale structures in the microwave background radiation. This then leads us to a whole new area of study which is whether or not all the matter in the Universe really is in the form of ordinary matter.

One of the most embarrassing things about astronomy and cosmology is that we are not really certain what form most of the mass in the Universe takes. It really is rather embarrassing to have to admit this only in the middle of the fourth chapter of this book. The way this comes about is as follows. We can measure the masses of galaxies and clusters of galaxies using essentially the same procedure we went through when we measured the mass of the black hole in the centre of NGC4151 (Chapter 3). We balance the centrifugal force acting on the matter by the gravitational force due to the system as a whole and we can find the mass of the whole system. When we carry out this type of calculation for giant spiral galaxies (Figure 63), we find that the light of the galaxy falls off very rapidly as we go out through a galaxy but the amount of mass falls off much more slowly. This means that there must be lots of mass in the outer regions of the galaxies which emits very little light. Such so-called dark matter is also found in the giant clusters of galaxies such as the Pavo cluster of galaxies which we showed in Chapter 1 (Figure 12). By measuring the velocities of many of the galaxies in the cluster, we can estimate its total mass and that turns out to exceed by a factor of about 10 to 20 what we would have guessed simply from the light of the galaxies in the cluster. It seems that there may be a lot of dark matter in the Universe on even larger scales because dynamical estimates of the amount of mass present provides more evidence that there must be much more matter present than we can account for simply from the starlight of galaxies. This is the famous **Dark Matter** problem. The key questions are - how much dark matter is there in the Universe and what sort of matter is it?

In many ways the first question is easier. The total mass density of the Universe is a key question for the whole of cosmology. We are going to show in the last chapter that this density determines whether the Universe will expand forever or whether it will recollapse to a Big Crunch. There is a critical density which just separates the behaviour of the models which expand forever from those which collapse back to the Big Crunch and that density is what we call the **critical density** of the Universe. We will call this critical density $\rho_{\rm crit}$. For our present purposes, the question is how much dark matter there is relative to the mass density of visible matter and what these densities are relative to the critical density. These are among the most important and difficult problems of modern observational cosmology. The amount of matter which is present in the visible images of galaxies corresponds to only about 1 to 2% of the critical density. In other words, we are confident

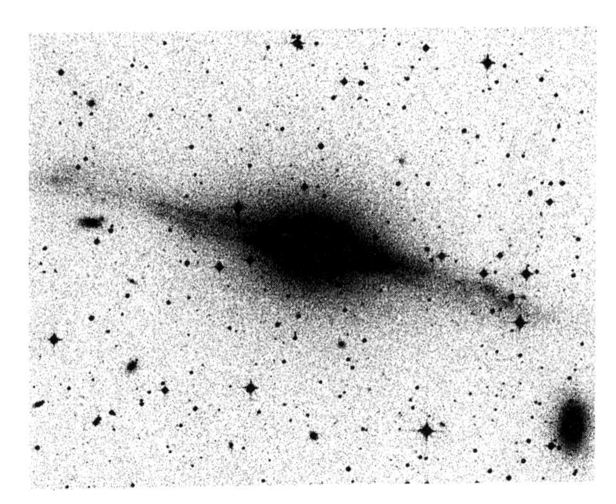

Figure 63. The edge-on giant spiral galaxy NGC5084 which is known to contain about ten times as much dark matter as visible matter.

that that this form of matter cannot "close" or stop the expansion of the Universe. Most estimates of the amount of dark matter relative to visible matter suggest that there is about 10 times more mass in the dark matter. If this result applies universally, we could account for about 10 to 20% of the critical density. What we don't know is whether or not, when we go to the largest scales, the density becomes equal to the critical density. There is certainly no reason why our Universe could not have the critical density although equally the Universe could have a relatively low density. Neither of these statements would be in contradiction with anything we know about the Universe. We will come back to this point in the last chapter.

Now, what could the dark matter be? Here is a bullet list of some possible forms for the dark matter.

- Interstellar planets
- Brown dwarfs
- Very low mass stars
- Massive neutrinos
- Unknown weakly interacting particles
- Little black holes
- Big black holes
- Supermassive black holes
- Abandoned spaceships
- Building bricks
- Old newspapers

We start with astrophysically very respectable things like very low mass stars (or brown dwarfs), black holes, supermassive black holes and then go on to some of the exotic types of particles discussed by the particle physicists and then onto really exotic things like spaceships, bricks, copies of old newspapers and so on. Now, some of the items on this list may look rather silly but they are there to make an important point. The dark matter could be in any of these forms and we would not be able to disprove that hypothesis, even the copies of old newspapers could make up the dark matter we know is present in the Universe. We only know if forms of matter are present in the Universe if they emit radiation which we can detect or if they absorb radiation from background objects. We can imagine lots of types of object in the Universe which would be very dark and the question is whether or not these make a significant contribution to the mass density of the Universe. In the Hot Big Bang model of the Universe, we can exclude some of these possibilities by astrophysical arguments so that I doubt if there was a reporter present at the Big Bang but you never know with reporters! You will notice that there are some remarkable forms of matter on this list. Some of these forms of dark matter must be important at some level in the Universe - for example, brown dwarfs, rocks and black holes must be present at some kind of population density in the Universe but we have no way at the moment of knowing if they could contribute as much mass as is needed, for example, to bind the giant clusters of galaxies. What has excited a great deal of interest among the particle physicists is the possibility that the dark matter may consist of exotic particles which would be very difficult to detect by existing accelerator experiments. We will have a lot more to say about these ideas in the last chapter.

What we want to do in this chapter is to look at the impact of dark matter upon the formation of galaxies. In particular, we want to consider some of the more exotic forms of dark matter because they may well turn out to be rather important in resolving the problems of forming galaxies which we identified above - specifically, the need to reduce the fluctuations in the microwave background radiation to an acceptable level. Suppose there exists in the Universe lots of exotic dark matter which at the present day only interacts with matter through its gravitational influence rather than through other forces such as electromagnetism or the weak and strong forces of particle physics. These types of particles are predicted by current theories of elementary particles. Suppose also that the dark matter is dynamically dominant in the sense that it contains much more mass than is present in the ordinary matter.

The clue to understanding what happens is to remember that, when the scale factor of the Universe is less than about 1/1000, the ordinary matter is very strongly coupled to the radiation and this makes it very difficult for this ordinary matter to collapse to form galaxies and clusters of galaxies before the matter and radiation decouple. However, the dark matter doesn't interact with radiation at all at these epochs. Therefore, perturbations in the dark matter can grow perfectly happily according to the relations we have derived above while the ordinary matter is prevented from collapsing by its strong interaction with the radiation prior to the epoch when the hydrogen recombines. Thus, we can imagine perturbations in the dark matter to have developed to quite finite amplitudes but we have no knowledge of this because these are not coupled to the ordinary matter. When the hydrogen recombines, the ordinary matter suddenly becomes neutral and is no longer tied to the radiation. The ordinary matter is free to fall into the perturbations already present in the dark matter and we can show that the perturbations in the ordinary matter very quickly grow to have the same amplitude as those in the dark matter. The attraction of this picture is that, at the time when the temperature fluctuations are imprinted on the microwave background radiation, the fluctuations in the ordinary matter are still very much smaller than they are in the dark matter, in particular, they are much smaller than the perturbations in the simple model we described above. This is a simple picture of how it is possible to obtain very small fluctuations in the matter which is coupled to the background radiation while the fluctuations in the dark matter are at the same time sufficiently large to ensure that galaxies form by now.

This is the currently favoured model for the origin of the large scale structure of the Universe but it is far from free from problems. Figure 64 shows some simulations of how we expect the galaxies and larger scale structures to form in two different versions of the dark matter theories of the origin of structure in the Universe. In one example, the case of **Cold Dark Matter** is considered - by this, we mean that the dark matter particles are very massive and decoupled from the ordinary matter very early in the Universe. One of the very intriguing possibilities is that the Cold Dark Matter might be in the form of some of the remarkable particles proposed by the most recent theories of elementary particles - for example, in supersymmetric theories, every particle has a supersymmetric partner and these would interact very weakly with ordinary matter. In this scenario, small scale structures survive to the epoch when the universal plasma recombines and the galaxies are built up by a process of hierarchical clustering under gravity. An alternative version is the **Hot Dark Matter** theory, the simplest form of which is to suppose that the electron neutrino has a finite rest mass of about 10 eV. This is a remarkably interesting mass because electron neutrinos are in equilibrium with matter when they come out of the Hot Big Bang and, if they had this rest mass, there would be enough mass to close the Universe. This picture is rather different from the Cold Dark Matter picture because it turns out that all the small scale structure is wiped out long before the epoch of recombination. Therefore, in this picture, the largest scale structures form first and then the galaxies form by fragmentation and condensation of the large scale structures. Remarkably, the two simplest dark matter models result in completely different pictures of how the galaxies formed. In the Cold Dark Matter picture, the galaxies are built up from smaller building blocks - in the Hot Dark Matter picture, the galaxies form by the fragmentation of larger scale structures.

These models have been the subject of detailed computer simulations and typical results of the modelling are shown in Figure 64. You will notice that they are quite successful in accounting for the fact that we observe large scale structures now. However, according to the experts, there are problems with both of the models. In the case of the Cold Dark Matter model, it is very difficult to account for the enormous voids, walls and stringy structures seen in the Harvard-CfA survey (Figure 14). This can be understood from the physical point of view because the basic process of formation of structure is hierarchical clustering which tends to wipe out stringy structures. On the other hand, the experts claim that the Hot Dark Matter picture produces too much structure - this is because, the matter cools on falling into the giant gravitational potential wells created by the dark matter and all the galaxies form in very thin sheets. Clearly, something is not quite right with the existing models.

One of the most important recent results of the experiments from the LEP experiment at CERN is that we can constraint some of the possible forms of dark matter. For example, the recent results on determining the number of neutrino species is of the greatest interest because it tells us that there are only three forms of neutrino. If the W and Z bosons were able to decay into any of these exotic particles, their presence would have already been detected in the LEP experiment if they had masses up to about 30 GeV. This can become a very technical story but the important point to remember is that there is now a very close relation between

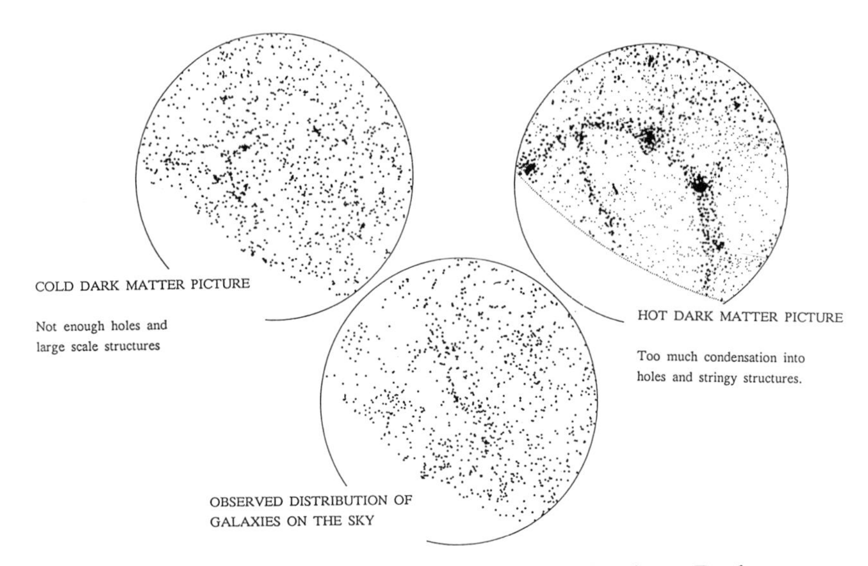

Figure 64. Recent simulations by Carlos Frenk at Durham University of the expectations of (a) the Cold Dark Matter and (b) the Hot Dark Matter models of the origin of the large scale structure of the Universe compared with the observations (c). The simple Cold Dark Matter model does not produce sufficient large scale structure in the form of voids and filaments of galaxies whereas the simple Hot Dark Matter model produces too much structure.

the physics of how galaxies might have formed and elementary particle physics.

There is, however, still a big problem. Although the dark matter helps, it does not get round the fundamental problem that we have to put in the correct spectrum of perturbations at the beginning if we are to make the galaxies, clusters of galaxies and larger scale structures by now. We have not been able to eliminate the problem of having to put in what we call an initial perturbation spectrum with finite amplitude perturbations in the beginning of the model. This is very worrisome because it means that we will only get out of our model what we put in at the beginning. This does not have the type of explanatory value one would look for in a good theory of galaxy formation. This is only the first of a number of features which, according to the classical theory, have to be built into our picture of the early Universe if we are to create the Universe as we know it by now. We will encounter three other problems of this sort in Chapter 5.

What we need is more direct observational evidence about how the galaxies and the large scale structure of the Universe came about. We cannot yet carry out these studies for ordinary galaxies but we are now able to study certain classes of luminous object back to times when the Universe was squashed by a factor of about 5 to 6 as compared to its present size, i.e. $R \approx 0.2$. This is not quite as spectacular as the factor of 1000 which takes us back to the epoch of recombination but it corresponds to looking back to when the Universe was between about one-fifth and one-tenth of its present age.

What do we see? The most extensive work has been carried out on very active galaxies, in particular, the quasars. You will recall that these are the most luminous objects in the Universe and therefore they can be observed at much greater distances than ordinary galaxies. When we survey very large numbers of distant quasars we find that they were very much more active in the past than they are now. If we plot the probability of finding powerful quasars as the Universe grows older, we find that when the Universe was about one-quarter of its present age, there was much more quasar activity then than there is now. What this means is that the process of galaxy formation must have been such that these galaxies had time to grow very massive black holes in their nuclei by the time the Universe was only about a one-quarter of its present age. That is a powerful constraint upon the formation and evolution of galaxies in the Universe. The only problem with using quasars in these studies is that their astrophysics is much more poorly understood than that of stars and galaxies.

Another very exciting recent discovery has been that we can observe certain classes of galaxy which show the types of evolutionary effects which we expect must be important as they settle down into their present forms. When we look at some other classes of active galaxy, the radio galaxies, they seem to have much younger populations of stars than those nearby and, in addition, they have associated with them very large gas clouds. An intriguing question is what the evolutionary status of these clouds is. Are these bits of the galaxy which are forming or are they simply gas clouds in the vicinity of the galaxy which are being excited by the activity going on in the active nucleus? We do not know the answer to these questions, nor do we really understand the place of systems such as these in the scheme of galactic evolution. The exciting thing from the point of view of the future of astronomy is that we can now observe galaxies and active galaxies when the Universe was up to about one-tenth of its present age. This means that we can start putting the whole subject of the origin and evolution of galaxies, their environments and the large scale structure of the Universe on a real observational footing rather than it simply being the province of theoretical speculation. This is why instruments like the Hubble Space Telescope and the next generation of very large ground-based telescopes are so important for astronomy. We are now at the stage where we can just see the most luminous objects as they were when the Universe was very young. But these are exotic and atypical objects in the Universe. What we need to be able to do is to undertake the same types of study but for the ordinary stuff of the Universe.

These studies have very profound implications for cosmology because the number of real facts which we have about the Universe is quite limited. The more we can pin down directly by observation how the galaxies and large scale structure have evolved into their present state, the more secure our understanding of the evolution of galaxies and the large scale structure of the Universe will be. It may eventually be that we will be able to read off directly by observation how it was that galaxies and the large scale structure of the Universe came about. This may not be a practical proposition for many years to come but at least we know exactly what we would like to do observationally.

5

The Origin of the Universe

Finally, we have to grapple with the biggest question of the lot - the origin of the Universe itself. In the last chapter, we discovered that there are features of the Universe which must have their origins in its very earliest phases. In this chapter we are to look at more of these problems, again within the context of the Hot Big Bang model of the Universe. In the last chapter, we described the problems of making galaxies and I only sketched in the background against which this process takes place. I want to tell that story rather better now and we will find three other major problems besides the problem of making galaxies which must have their origins in the very early Universe.

We described briefly the dynamics of the Universe but let us now develop a better picture because there are many new features we have to build into the story. We established that the correct way of beginning the construction of models of the Universe is to consider models which are isotropic and uniformly expanding. What are the dynamics of these types of Universe? The fact that they are uniform and isotropic makes all the calculations very easy - you can think about the dynamics as a competition between the expansion which is dispersing all the matter in the Universe and the force of gravity which is trying to prevent that happening - i.e. to pull all the matter back together again.

It turns out that we can model the dynamics of the Universe exactly if we ask "What is the decelerating force acting upon a galaxy which sits at the edge of an expanding sphere which has density equal to the average density of the Universe?" Such a model is shown in Figure 65. This is

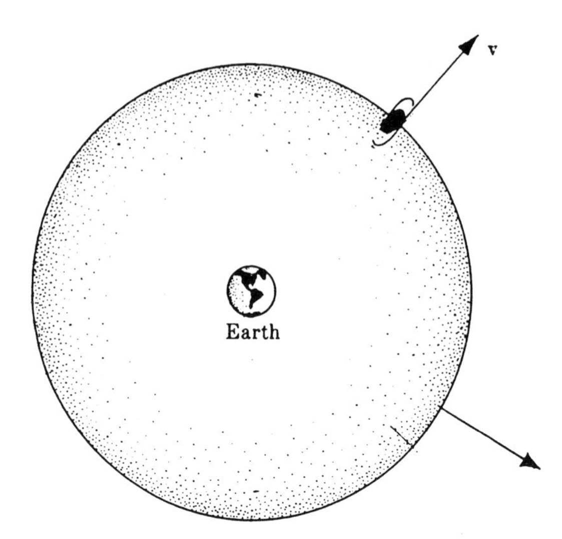

Figure 65. A simple model for working out the dynamics of an expanding Universe.

actually a very simple sum and it works because every observer in the Universe would carry out the same calculation. Remember that every observer regards the Universe as expanding spherically symmetrically about him or her because we only deal with uniformly expanding Universes (see Figure 62). This calculation results in the exact dynamics which come out of the complete theory of the expanding Universe according to Einstein's general theory of relativity. The reason for this is very interesting and is that, in the simple isotropic models of the Universe, local physics is the same as global physics - in other words, the properties of each bit of the Universe must also contain information about the global structure of the Universe.

The dynamics of the standard models of the Universe depend entirely upon the local density of matter. If the Universe is of high density, the gravitational deceleration acting upon each piece of matter can be sufficiently great to halt the expansion and the Universe will collapse back to a hot dense phase again - the Big Crunch. If the Universe is of low density, there may be insufficient mass in the Universe to stop it expanding to infinity. In this second case, the Universe ends up being of infinite size and it is still expanding with a finite velocity when it gets to infinity. Between these two extreme types of behaviour, there is what

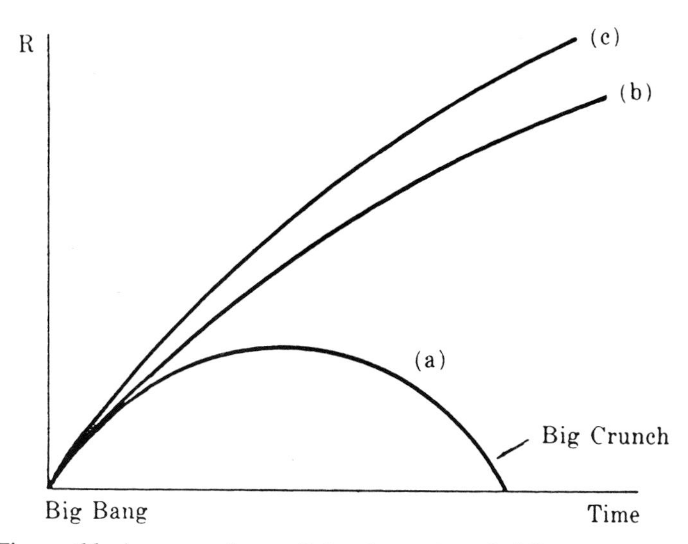

Figure 66. A comparison of the dynamics of different models of the expanding Universe. (a) A high density model which eventually collapses to a Big Crunch. (b) The critical model which has density ρ_{crit}. (c) A low density model in which the Universe expands to infinity and has finite velocity when it gets there.

is known as the **critical model**. This is the model which just expands to infinity and stops there. In other words, the velocity of expansion of the Universe goes to zero when the Universe is of infinite extent. The critical density is a very important concept in cosmology and we will come across it over and over again. As in Chapter 4, we will call it ρ_{crit}. The dynamics of these models of the Universe are shown in Figure 66 in which we describe the expansion of the Universe in terms of the **scale factor** R which simply describes the way in which the relative separation of any two points in the expanding Universe changes with time.

There is an exact analogy with the concept of escape velocity which is illustrated in Figure 67. A spaceship only escapes from the pull of Earth's gravity if it has a large enough velocity on take-off. If the velocity of the spaceship exceeds the escape velocity, it escapes to infinity. If it has less than the escape velocity, it does not escape from the pull of the Earth and falls back to Earth again. Between these extremes, the spaceship has exactly the escape velocity and it can only just escape to infinity and no more. These three types of behaviour correspond exactly

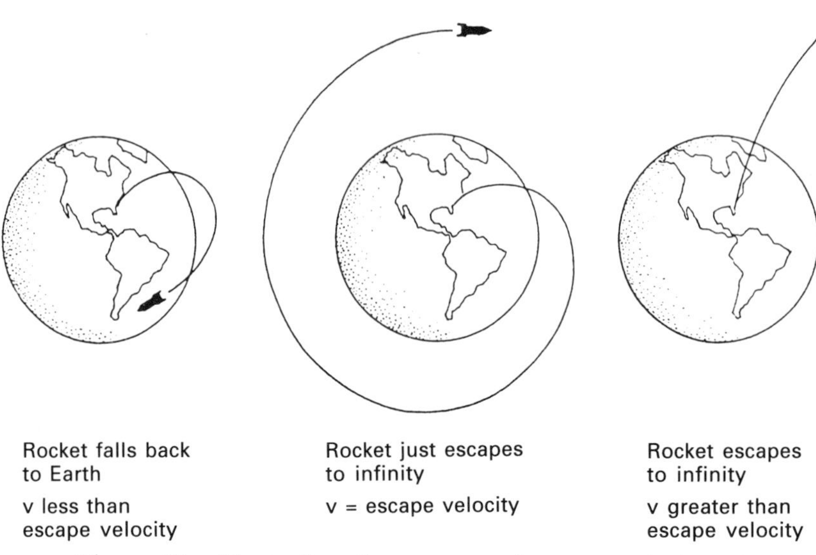

Figure 67. Illustrating the concept of escape velocity of a space rocket from the surface of the Earth.

to the dynamics of Universes which have density less than, greater than and equal to the critical density. In these simple models, the deceleration is entirely determined by the average mass density of the Universe.

There is one other interesting and important point about these models and that is that there is a very close relation between the geometry of the Universe and its average density. This is an aspect of the models of the Universe which we cannot recover from Newtonian theory but which is built into Einstein's relativistic theory of gravity, the general theory. What Einstein's theory tells us is that matter bends the geometry of space-time so that it is most unlikely that the global geometry of space is flat. However, there is a beautifully simple relation between the global geometry of the uniform isotropic Universes and their densities. If the density of the Universe exceeds the critical density, the spatial geometry is spherical. If the Universe is of low density, the geometry of space is hyperbolic and only in the case of the critical model is the geometry Euclidean. We need not worry too much about this locally because the radius of curvature of space is roughly the size of the Universe and so the angles of a triangle add up to 180° very precisely on a sheet of graph paper. If, however, our sheet of graph paper extended out to what we called the "edge of the Universe" in Chapter 1, then, the angles of a triangle with sides that big might no longer be equal to 180° - indeed, in general will not be. We will need this concept of the relation between geometry and the mass density of the Universe later at a crucial point in the discussion of the early Universe.

There is another remarkable thing about this story and that is the relation between the dynamics of the Universe and its mass density. Our simple picture shows us that the rate at which the Universe is decelerating is determined by the average density of matter. We can therefore try to do two things. First, we can try to measure the rate at which the Universe is decelerating and, second, we can measure how much mass there is in the Universe and see how well they match up - i.e. can we attribute the observed deceleration of the Universe to the mass which we know to be present? If we were able to show that the deceleration of the Universe is entirely caused by the amount of matter present, that would be an incredible validation of Einstein's general theory of relativity on the largest scales we can study in the Universe.

We can quantify this problem in the following way. First, we define a **density parameter** which is just the ratio of the actual density of the Universe ρ to its critical value $\rho_{\rm crit}$. We call this ratio $\Omega = \rho/\rho_{\rm crit}$. We can also define the deceleration of the Universe and this is conventionally measured by something called the **deceleration parameter** which is labelled q_0. Let us not go into the details of how the deceleration parameter is defined but it turns out that the deceleration parameter should be equal to exactly one half of the density parameter according to the standard models we have derived above. Therefore, the critical density for which $\Omega = 1$ corresponds to a value of the deceleration parameter $q_0 = \frac{1}{2}$.

You will remember that we discussed the density parameter briefly in the last chapter and all we really have are lower limits to its value. If we consider the visible matter in galaxies, we find a value of about $\Omega_{\rm gal} = 0.02$. If we include the dark matter, which we know must be present in large scale systems, the value of Ω increases by a factor of about 10 to $\Omega = 0.2$. There is some evidence that the value might be higher than this on the very largest scales but this is uncertain. What we can state with some confidence is that the Universe is probably within about a factor of 10 or so of its critical density. Measurements of the deceleration parameter q_0 are also very difficult because they essentially depend upon being able to measure the change in the velocity of expansion of the Universe between now and when the Universe was significantly younger than its present age. There are several classical tests by which this measurement can be attempted but the uncertainties in these estimates is still quite large. All that we can really state with some degree of confidence is that probably the deceleration parameter q_0 lies somewhere in the range 0 and 1. Thus, the probable range of values of the deceleration parameter spans both the critical value, $q_0 = \frac{1}{2}$ and low density models $q_0 \sim 0$. What this means is that, within about a factor of 10, the density and deceleration parameters agree with the standard theory and that is quite encouraging. If one is an extreme optimist, one might argue that, because they must be more or less the same, it is very likely that they are in fact the same - you must make up your own mind about that argument. I believe the argument has some substance but I would much prefer to have much improved estimates of both parameters.

This simple feature of the Universe, that its density is within about a factor of 10 of the critical density, leads to the first of the big problems. According to the standard models of the Universe, there is no reason why the density parameter should take any particular value at all. If we were to create Universes at random, there is no reason why the density parameter should be close to the critical value. You can imagine it being many billions of times greater or less than the critical value and it would not contradict any piece of basic physics. There is nothing in the physics of the models which tells us what it should be. The problem arises because if the value of the density parameter departed even very slightly from the critical value in the very early Universe, we can show that it would depart from the critical value now by an enormous factor.

The simplest way of illustrating this problem is to look again at our space rockets trying to escape from the Earth. Suppose the rocket starts off with a velocity which is just greater than the escape velocity. As the rocket travels further and further away from the Earth, the escape velocity at that point is less than that on the surface of the Earth because it is further away from the Earth. However, the ratio of velocity of the rocket to the local escape velocity gets bigger and bigger as we move further away from the Earth. We can look at the ratio of these velocities as the rocket travels further and further from the Earth and you can see that the divergence is enormous (Figure 68). It follows that, if we are very far from the Earth and we find that the rocket has more or less the escape velocity at that point, it must have started out with almost exactly the escape velocity with quite fantastic precision.

The Universe has exactly the same problem - unless it was set up with more or less exactly its own escape velocity at the beginning, it has no chance of ending up with an expansion velocity close to its escape velocity now. You can see what this means - the Universe must have

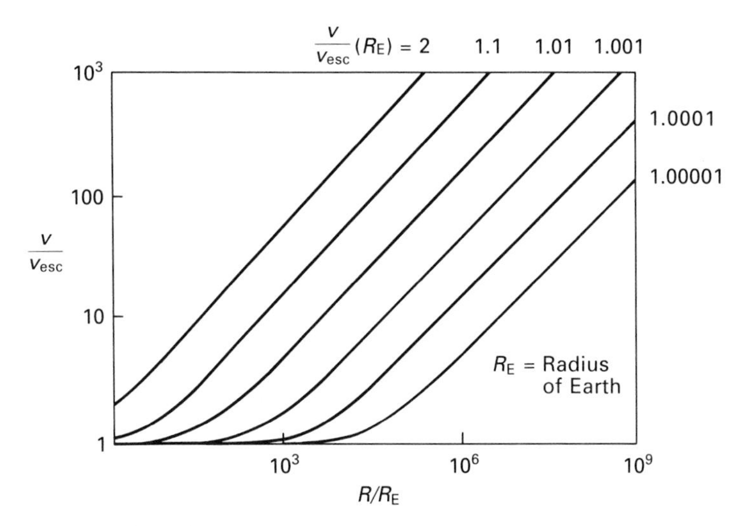

Figure 68. Illustrating how the ratio of the velocity of a space rocket to the local escape velocity changes as the rocket moves away from the Earth. The different curves show how the ratio changes for different assumed values of the initial escape velocity. Notice that, even if the rocket starts off with a velocity very close to the escape velocity, the ratio always diverges at large enough distances.

been set up in the beginning with almost exactly the critical density with quite amazing precision if it is to end up within a factor of 10 of the critical density now. This is what is called the fine-tuning problem of the Universe. Despite the fact that there is no obvious reason why our Universe should have the critical density, the Universe must have been set up that way very precisely in the beginning. The theorists argue very strongly that, because of the fine-tuning problem, the density parameter must be 1 now and that is a problem because we are about a factor of 5 or 10 short if we include known forms of dark matter.

Part of the problem stems from the fact that there is only one Universe we can study and that is the one we live in. We cannot go out and look at other Universes to see if they have done the same thing or not. Therefore, we cannot exclude the possibility that our Universe has these properties just by chance. There is an intriguing line of reasoning which states that there are only certain types of Universe in which life as we know it could possibly have formed. For example, there must have been enough time for biological systems to form, the stars must live long enough for life to form and so on. This is an approach which is called the **Anthropic principle** and it states in its simplest form that the Universe is as it is now because we are here to observe it! I will not continue with this line of reasoning any further because, personally, I would only adopt this approach as the very last resort if all other approaches to understanding the present properties of the Universe fail. I believe we are far from that position yet.

So far we have talked only about the matter. What about the radiation? We have already mentioned the fantastic isotropy of the microwave background radiation (Chapter 4). This means that this radiation fills all of space very uniformly. The Cosmic Background Explorer has measured not only the isotropy of the radiation but also the spectrum of the radiation and we displayed these very beautiful measurements in Figure 25. The little boxes show the observations and the line through the points is a theoretical line showing what is expected if the radiation has what is known as a **black-body spectrum**. You can see that the spectrum is an almost perfect black-body spectrum, certainly the most perfect I am aware of anywhere in nature.

Now what is the significance of this observation? Any body at a given temperature emits radiation simply because it is at that temperature. If you leave any object for a very long time in an enclosure at that same temperature as the object, eventually the radiation and the matter in the enclosure come to what we call thermodynamic equilibrium in which every emission is exactly balanced by a corresponding absorption. The radiation spectrum inside the enclosure then has a unique form which is determined only by the temperature of the body - it is the spectrum of black-body radiation. The fact that the microwave background radiation has such a perfect black-body spectrum tells us that at some stage all the radiation and matter in the Universe were in thermodynamic equilibrium.

There is one other beautiful result which comes out of the theory of the expanding Universe and that is that, when a black-body spectrum expands with the Universe, the radiation preserves its black-body spectrum but the temperature of the radiation is lower. What is more, the way in which the temperature decreases is very simple indeed. It is simply inversely proportional to the scale factor R. Now, the temperature of the microwave background radiation as measured by COBE is 2.736 K. So, if, for simplicity, we take the microwave background radiation to have a temperature of 2.75 K at the present time, when we squash the Universe by a factor of 2, the radiation temperature must have been greater by this same factor, i.e. the temperature was 5.5 K. As we squash the Universe backwards in time, the scale factor R becomes very small indeed and consequently the temperature of the radiation becomes very, very large.

We now understand why it was that the Universe became so hot in the past and, in particular, why the temperature of the radiation was about 4000 K when the Universe was squashed by a factor of 1500. You will recall that we said that at about that time all the hydrogen in the Universe was split up into its constituent protons and electrons and this provided very strong coupling between the matter and the radiation. At earlier times, the temperature was higher and the coupling between the matter and radiation was even stronger so that early enough the matter and the radiation were kept at exactly the same temperature. Thus, in the early Universe, the matter and radiation were more or less in thermodynamic equilibrium and this is why the radiation has such a perfect black-body spectrum.

One interesting feature of this model is that, in the early Universe, the radiation has much more mass than the matter. What I mean by this is as follows. Einstein tells us that $E = mc^2$ and hence the energy E present in radiation has a certain mass m. It turns out that, at those times when the scale factor was less than about 1/10 000, the dynamics of the Universe were determined by the mass in the radiation rather than that in ordinary matter. We call these early phases of the Universe the **radiation-dominated epochs**.

As we go further back in time, the next critical stage occurs when the Universe was squashed by a factor of about 1 000 000 000. You may say, "Doesn't the Universe become quite fantastically dense when we squash it by this huge factor?" Well, the answer is perhaps slightly surprising in that it is "No!" The reason is that the average density of matter in the Universe is really very, very low at the present time. In fact, when the Universe is squashed by a factor of 1 000 000 000, the average density of the Universe is still only roughly that of air in a typical room. Now, we believe we understand the behaviour of matter at these densities rather well, the only difference being that all the matter and radiation are now at a temperature of about 3 000 000 000 K. This is a very interesting temperature because it is so high that the black-body spectrum is shifted to γ-ray energies and these γ-rays are energetic enough to dissociate the nuclei of atoms into their constituent elementary particles. In other words, the nuclei of any chemical elements which happened to be around at that time were broken up into protons and neutrons by the thermal radiation. At even earlier times, there were no longer any nuclei of atoms present - the Universe becomes a sea of elementary particles. Collisions between these particles can create the whole zoo of elementary particles when the temperature becomes high enough.

What guarantee do we have that this really happened? Until 25 years ago, this was simply theoretical speculation but detailed calculations have now been made of what we might expect to survive from these very early phases as the Universe cools down. What is done is to evolve model Universes from states of very high density and temperature, including all the very best physics about the interactions between all the elementary particles we know of. The results of some of these classic calculations by Robert Wagoner, William Fowler and Fred Hoyle at Caltech show how the abundances of various light particles and nuclei evolve in the early Universe through the critical time when the Universe was only a few minutes old (Figure 69). These are really extraordinary calculations.

Let's understand what is happening in Figure 69. In the beginning, we have only the constituents of nuclei - the protons and the neutrons as well as radiation and other particles which we know must be around because the system began in thermal equilibrium - things like neutrinos, muons and so on. As the Universe cools, we find that the protons and neutrons begin to combine to form the simplest compound nucleus - deuterium (^{2}H) which contains a proton and a neutron. Then, as the cooling proceeds further, helium (^{4}He) and traces of other elements such as the isotopes of helium (^{3}He) and perhaps a little lithium (^{7}Li) are created through collisions among the deuterium nuclei. What is striking is that we are not able to make any of the heavy elements such as carbon and oxygen - the reason for this is that there is no stable species with four protons and four neutrons - that would be Beryllium-8 (^{8}Be) but it is totally unstable.

The spectacular result of these calculations is that we expect large amounts of helium (^{4}He) to be produced as well as small but significant amounts of deuterium (^{2}H) and helium-3 (^{3}He) and traces of lithium (^{7}Li). Now, it has always been a big problem to understand why, wherever we look in the Universe, helium always seems to be present with an abundance of about 25% by mass. The remarkable result is that, for any reasonable value of average density of ordinary matter in the Universe now, we predict that about 23 to 24% of helium by mass will be created by nuclear reactions in the first three minutes of the Big Bang. We do not know of any other way of making this large amount of helium by synthesis inside stars. The Hot Big Bang solves this dilemma.

The other wonderful realisation has been that the other light elements

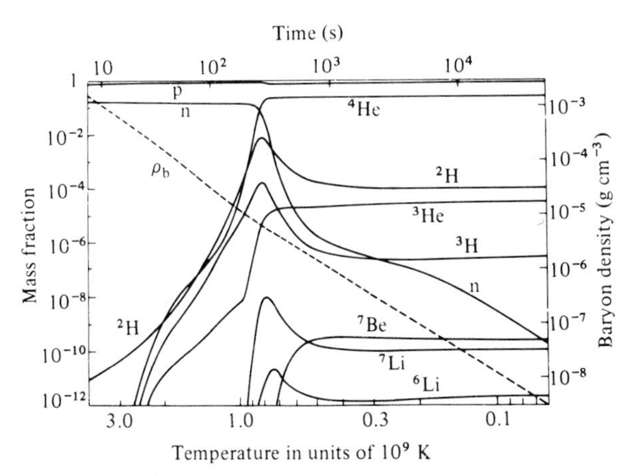

Figure 69. A diagram showing how the abundances of the chemical elements are expected to change during the first 30 minutes of the Hot Big Bang. The time from the Big Bang is shown along the top of the diagram and the corresponding temperature along the bottom. The diagram shows how the mass fractions of the different elements build up as the Universe cools down.

are also important tracers of the properties of our Universe and, in particular, deuterium is of special importance. Deuterium is the heavy isotope of hydrogen and it cannot be created in stars by nucleosynthesis. It is such a fragile nucleus that, as soon as it is produced in any significant amount in the centres of stars, it is immediately destroyed by the radiation present at the very high temperatures at which it was created. In other words, we never expect stars to be able to make any deuterium at all. In the Hot Big Bang, significant amounts of deuterium are produced. The reason it can survive is that the Universe expands so rapidly in its early phases that there is not time to destroy it all by γ-rays or to convert it by further reactions into helium. It turns out that the amount of deuterium present is a very sensitive probe of the density of ordinary matter in the Universe. If the Universe were of high density, then there would be sufficient collisions between deuterium and other nuclei to convert almost all of the deuterium into helium. If, however, the Universe is of low density, there is not enough time for all the deuterium to be converted into helium and that leaves some deuterium left over which we can observe now.

Now, notice the interesting way in which this argument proceeds - the more deuterium created in the Hot Big Bang, then the lower the average density of ordinary matter in the Universe. Since we only know of ways of destroying deuterium, the present abundance of deuterium sets a rather firm upper limit to the density of ordinary matter in the Universe. If the density were any greater, we would not produce enough deuterium. The reason that this is such an exciting result is that it tells us that the density of ordinary matter in the Universe must be less than about one-tenth of the critical density. In fact, it turns out that we can explain the observed abundances of helium-3 and lithium as well for a single value of the present mass density of the Universe (Figure 70). This means that we have no problem in accounting for the fact that the material which makes up the visible matter in the Universe corresponds to about 2% of the critical density. However, we know that there must be dark matter present which is about 10 times more abundant - can this all be ordinary matter or not? The answer is that we do not really know. My own belief is that we can probably reconcile the limits to the density of ordinary matter in the Universe with the amount of dark matter we need in the outer regions of galaxies and in clusters of galaxies but it is a close run thing.

What these calculations tell us is that we cannot attain the critical density with ordinary matter. This is a crucial point. If the Universe had the critical density in ordinary matter, very low abundances of deuterium would be produced in the Hot Big Bang. If our Universe really has the critical density, and we have argued that this may be the only stable value it can have, then most of the matter in the Universe cannot be ordinary matter - this is what is exciting the particle physicists. The dark matter may well be in some exotic form such as those predicted by currently favoured theories of elementary particles. Examples of these possibilities include the hypothetical supersymmetric partner of the photon, the photino, or possibly the gravitino, or axions, and so on. This is why the subject of the very early Universe has been taken over with great enthusiasm by the particle physicists who regard it as a laboratory for testing out their theories of elementary particles. I have a lot of sympathy for them because it is the only way they can obtain high enough temperatures to test their theories!

Notice that this line of reasoning completes the third of the three

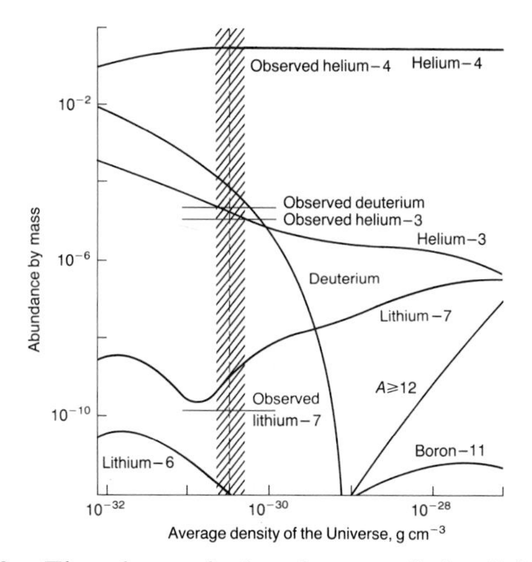

Figure 70. The observed abundances of the light elements compared with the predictions of difference standard world models. For a range of low-density world models, the Hot Big Bang models can account for the observed abundances of helium, deuterium, helium-3 and lithium.

independent arguments which support the Hot Big Bang model. These are:

1 The Hubble expansion of the Universe.
2 The isotropy and thermal spectrum of the microwave background radiation.
3 The abundances of the light elements.

These three pieces of evidence are independent and all find a natural explanation in the standard Hot Big Bang model of the Universe.

We still have two more problems to discuss. The first of these can be understood by considering what happens when we go to yet higher and higher temperatures in the early Universe. One of the great discoveries of particle physics of the 1930s was that for every type of particle, there must be an antiparticle which is a mirror image of that particle. So, for example, we know that, in high energy collisions between high energy γ-rays and ordinary matter, large numbers of electrons and their mirror-image particles the positive electrons, or positrons, are created. Figure 71 represents a particle physics experiment in which this process has

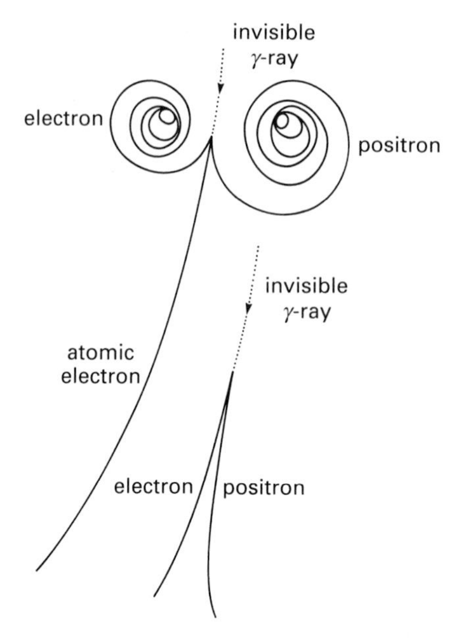

Figure 71. An example of electron-positron pair production taking place in a particle physics experiment. The electrons and positrons have opposite charges and so are bent in opposite directions by a magnetic field.

been captured. The electrons are negatively charged particles and the positrons are identical but with opposite charges and so their paths are bent in opposite directions by a magnetic field. In fact, all the properties of antiparticles are reversed as compared with particles. In exactly the same way, protons have a positive electric charge and their mirror-image particles, the antiprotons, have negative charge.

Now, if we take the Universe back to early enough times, the thermal background radiation attains γ-ray energies and these γ-rays can collide, producing equal numbers of particles and antiparticles, provided, of course, that they have enough energy to create the particle-antiparticle pair. This is just Einstein's mass-energy relation $E = mc^2$ applied to the collision - in other words, the photons have to have enough energy to create the masses of both the particle and the antiparticle. This results in a rather curious situation because we know that locally our Universe is only made out of matter, rather than being an equal mixture of matter and antimatter. There must be very little antimatter present or we would have recognised its presence by the γ-rays which would be emitted when it annihilates with matter.

We can work out how many γ-rays there are in the early Universe relative to the number of particles of matter and it turns out that there are about 100 000 000 to 1 000 000 000 γ-rays for every proton. This results in a rather curious paradox because, when the temperature gets so high that we can create protons and antiprotons from the high energy γ-rays, the Universe will suddenly be flooded with millions of protons and antiprotons. When we now run the clocks forward from that very hot early phase, we have to start off with, say 1 000 000 001 protons and 1 000 000 000 antiprotons. As the Universe cools down, the 1 000 000 000 protons annihilate with the 1 000 000 000 antiprotons and produce γ-rays, leaving only one proton left over which becomes the Universe we know and love. You can now understand the problem - we cannot start off with a Universe which is completely symmetric between matter and antimatter or else almost everything would annihilate. We would end up with a very small amount of matter and antimatter in the Universe, far less than the amount we observe. Hence, in the classical picture, we have to build in a slight asymmetry between the numbers of protons and the antiprotons or more generally, particle and antiparticles, in the very early Universe.

The final problem concerns the remarkable isotropy of the Universe. In its simplest form, we can ask, "How was it that one bit of Universe knew that it had to end up looking exactly the same as another bit of Universe which is very distant from it now?" According to conventional thinking, if you want one bit of Universe to have the same properties as another bit, the very least you have to be able to do is to ensure that the two regions can communicate with each other, since otherwise they will not "know" that they have to be the same. Now, the fastest that this information can travel is the speed of light. If the Universe is expanding, this turns out to present a nasty problem, because there is only a finite amount of time since the beginning of the expansion.

Let us illustrate how this results in a major problem by studying what happens on a simple space-time diagram of our Universe. Now, in the conventional version of this diagram, we plot time up the vertical axis and distance along the horizontal axis. We plot the lines corresponding to the paths of galaxies partaking in the expansion of the Universe (Figure 72). Now, when we look at the Universe, we look at it as it was in the past and this information comes to us at the velocity of light. So, we can only receive information from along a cone centred on the Earth. I have shown this as a past light cone centred on our Galaxy. Now,

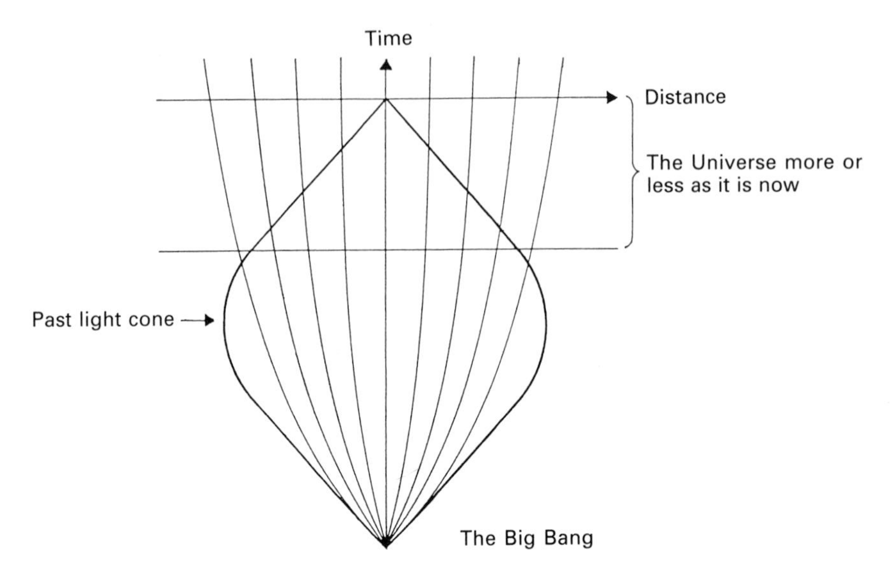

Figure 72. A simple space-time diagram of our Universe showing the bending of the past light cone towards the singularity at the Big Bang.

unfortunately, it cannot remain a simple cone too far back, or else it won't go through the Big Bang. In fact, the past light cone has to bend round at large distances as I have shown on this diagram.

Now, this is where the cosmologists perform a nice trick - let's distort our coordinates so that the past light cone remains a genuine cone but the paths of the galaxies are distorted. The Big Bang then occurs over a sheet rather than at a point. This enables us to understand the problem rather better (Figure 73). We know that the microwave background radiation is remarkably isotropic and that it was last scattered when the scale factor was about $1/1000$. In Figure 73, I show schematically two patches of sky in opposite directions from which the microwave background radiation was emitted and attached to each of these are their own past light cones. You can see that they do not overlap even if we go right back to the Big Bang. This is the last problem - how was it that the Universe knew that it had to have the same properties everywhere, even in very remote regions which could not communicate according to our simple construction? The only way of explaining this problem in the classical picture is to assume that that is how the Universe was set up.

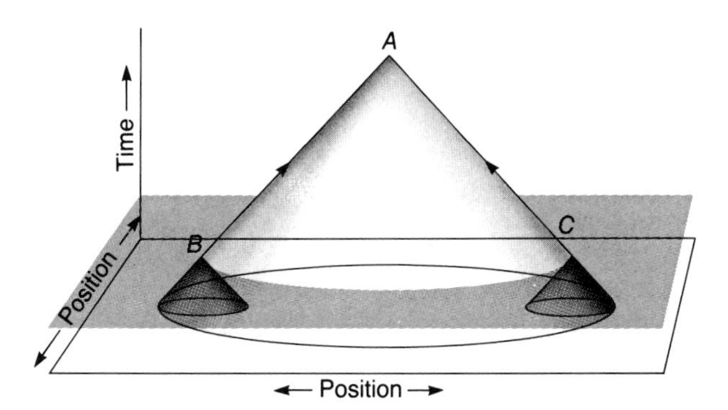

Figure 73. Illustrating the horizon problem in the expanding Universe. The system of coordinates has been distorted so that the past light cones are true cones as shown. To achieve this the trajectories of galaxies are distorted and the Big Bang is stretched out into a plane. We observe distant regions of the Universe to be the same but there was not time in the standard picture of the Universe for these widely separated regions to know that they had to be the same.

We have now completed our catalogue of the four great problems of cosmology. Just to recap what they are:

1 Why is the Universe so isotropic?
2 Why is the Universe so close to its critical density?
3 Why was the Universe slightly asymmetric with respect to matter and antimatter in its early phases?
4 What is the origin of the fluctuations from which galaxies formed?

In the classical picture of the Hot Big Bang, these four problems are solved by assuming that the Universe starts off with the correct initial conditions to begin with. In other words, we have to postulate that the Universe was set up with density close to the critical density, that it was isotropic, that there were fluctuations present out of which the galaxies were made and we have to assume that there was a slight matter-antimatter asymmetry, all as initial conditions. This may be correct but, scientifically, it is not a very pleasant picture because it has no explanatory value - we only get out the end what we put in at the beginning.

Fortunately, we now have some clues as to how we may be able to resolve these problems by looking at the impact of particle physics upon the evolution of the very, very early Universe. Many of these ideas are very new and would take us far into the complexities of modern particle physics. Let me be very selective and give you a flavour for how one or two of these ideas give me hope that we will eventually be able to tie many of these problems together into a unified physical picture of the very early Universe. At the moment, these ideas must be regarded as speculative but they contain within them new concepts which are, in principle, testable in the very early Universe. This process is no different from the way in which Copernicus discovered that the planets orbit the Sun, or Newton derived his law of gravity from Kepler's laws of planetary motion. The modern day physicist uses the early Universe as a laboratory within which to test theories of elementary particles.

Let us discuss first of all the very early dynamics of the Universe. To cut a very long story short, there is one very attractive way of resolving two of the problems on our list - these are the isotropy of the Universe and the fact that it is close to the critical density. The germ of this idea comes from some of the earliest applications of Einstein's general theory of relativity to the Universe. Among the formal solutions are models in which the Universe expands exponentially in its early stages. In other words, the Universe doubles in size in successive equal time intervals. You can imagine the Universe beginning very tiny and very, very dense indeed and then expanding at a quite extraordinary speed under the influence of this exponential expansion. This picture is called the **inflationary** picture of the early Universe. This process has some remarkable features. One very obvious one is that we can start with regions of the Universe which are very close together indeed and which consequently can communicate with each other, and then let them expand apart at an exponential rate so that they rapidly move to such distances that they can no longer communicate with each other by light signals. This can solve problem Number 1.

Now let us look at the geometry of the Universe in this type of inflationary expansion. If the Universe expands by this enormous factor exponentially, it is just like blowing up a balloon. The curvature of space may be very complex in the beginning but, as the balloon blows up, the curvature of the surface becomes flatter and flatter. In fact, if the expansion is large enough, we end up with essentially flat Euclidean geometry. Now, as we explained earlier, in cosmology the geometry of space-time and its matter density are very closely related and so, if we can now transform over from this exponentially expanding Universe to our ordinary Universe by some suitable means, we end up with a Universe which has Euclidean spatial sections and consequently must have density very close indeed to the critical density. This comes about because, as we mentioned earlier, according to the classical theory of the expanding Universe, the critical model has flat Euclidean spatial geometry. These are two remarkable results of the idea that, in the very early stages, the Universe went through a period of very rapid expansion - what is known as the **Inflationary Universe**. What is the price we have to pay to make this work? The answer is that we have had to introduce a special type of physical force in the very early Universe. We can show very simply that this type of behaviour only occurs if the matter has a very peculiar equation of state. It is what is known as a **negative energy equation of state**. An ordinary gas has an equation of state in which the greater the density of the gas at a given temperature, the greater the pressure. In a negative energy equation of state, the opposite happens in that the pressure is proportional to **minus** the density of the matter. In fact it is not a pressure at all - it is the exactly the opposite and should be called a tension. The reason that this idea is taken seriously is that the particle physicists have found forces in elementary interactions which have exactly these properties. The simplest of these forces is found in the theory of the unified electro-weak interaction and is needed in order to give the particles mass. The important point is that there is now genuine physical motivation for the introduction of forces with these strange properties at very high energies. As yet there is not a proper physical realisation of the forces which could result in the inflationary expansion of the early Universe but these ideas look encouraging.

What about the third problem concerning the asymmetry between matter and antimatter. Again, there are powerful clues from particle physics. We know that there is a built-in asymmetry between matter and antimatter when we look at the results of accelerator experiments. In particular, a neutral particle known as the K^0 meson, should decay symmetrically into equal numbers of particles and antiparticles but in fact doesn't. It has a very slight preference for matter over antimatter. We can built this idea into the early Universe and the theorists tell us that we can at least get roughly the right asymmetry between matter and antimatter that we observe in the Universe. This looks encouraging too.

The fourth problem about the origin of the fluctuations from which we make galaxies is even more technical than the first three and we simply note that the very early Universe does not simply cool and grow cold. There are a number of phase changes as the Universe expands. Whenever these changes of state occur, just as when water freezes or boils, there is the possibility of creating large fluctuations. One possibility is that, out of these phase changes or similar phenomena, we may find the fluctuations we need. Another possibility is that the fluctuations develop out of quantum fluctuations in the very early Universe.

I hope the general trend of the argument is clear. Nobody knows the correct solution to the four great problems but there is much more than a glimmer of hope that particle physics combined with astrophysics will enable us to obtain a much deeper understanding of how they can be solved. It would be a quite remarkable synthesis if the very, very large and the very, very small come together in the very early Universe - but that is the current goal of much cosmological research in this field.

Have I answered the question about what the origin of the Universe is? Clearly, I haven't. When we consider these types of problems, we are not talking about the Universe when it was a second or even a millionth of a second old. We are considering densities which exceed anything that we have experience of in our laboratories. There is no other way we know of at present of studying the very extreme physical conditions of the very early Universe. But there is lots of new physics and astrophysics to be understood. It is in this interaction between the new physics and the evolving Universe as we know it that lie perhaps the greatest of all challenges for the modern cosmologist.

Further Reading

There are many excellent introductory books on modern astronomy and cosmology. The following are among the books I refer to most:

The New Physics, ed. P. C. W. Davies, 1989. Cambridge University Press. This book contains a number of articles on general relativity, cosmology and astrophysics roughly at the level of *Scientific American.*

The Cambridge Atlas of Astronomy, ed. J. Audouze and G. Israel, 1989. Cambridge University Press. This book is profusely illustrated with many excellent images of all classes of objects in the Universe. The level is non-technical.

Alice and the Space Telescope by M. S. Longair, 1989. Johns Hopkins University Press. This book was written as a celebration of the science which can be undertaken by the Hubble Space Telescope. It contains a series of brief essays about many topics in modern astronomy.

Fundamental Astronomy by H. Karttunen, P. Kroger, H. Oja, M. Poutanen and K. J. Donner, 1987. Springer-Verlag. This is a beautiful modern textbook on basic astronomy. It provides an excellent introduction at a slightly more technical level to many of the topics in this book.

Physics of the Early Universe eds. J. A. Peacock, A. F. Heavens and A. T. Davies, 1990. SUSSP Publications. If you are very brave and need a graduate level introduction to the problems of galaxy formation and the origin of the Universe, this is the book for you. You should not be afraid of quantum field theory and elementary particle physics before you open this book.

Universe by M. Rowan-Robinson, 1990. Longman. This very nice book contains up-to-date images of many of the objects I have referred to here.

Picture Acknowledgements

1 Courtesy of NASA and the Space Telescope Science Institute (1990); *3* Courtesy of the Mount Palomar Observatory; *4* Courtesy of the Lund Observatory; *6, 7, 34* Courtesy of D. Malin and the Anglo-Australian Observatory; *2, 8, 12, 45* Courtesy of the Royal Observatory, Edinburgh; *9, 10* Courtesy of the Royal Greenwich Observatory; *11* Courtesy of H. Arp from *Optical Jets in Galaxies* (1981), ESA Publications; *13* Courtesy of P. J. E. Peebles from *Astron. J.* (1977); *14* Courtesy of M. Geller, copyright 1990 Smithsonian Astrophysical Observatory; *17* After Giacconi, Gursky and van Speybroeck, from *Ann. Rev. Astr. Astrophys.* (1968); *19* From S. van den Bergh, in *Astrophys. J.* (1975); *20* Courtesy of Dr F. Seward and the Harvard-Smithsonian Center for Astrophysics; *21* Courtesy of NASA; *22* Courtesy of Dr K. Bennett and the European Space Agency; *23* Courtesy of NASA, the Jet Propulsion Laboratory and the Rutherford-Appleton Laboratory; *24, 25, 61* Courtesy of NASA/Goddard Space Flight Center (1990); *26* Courtesy of Dr G. Haslam, Max-Planck Institute for Radio Astronomy, Bonn; *29* Courtesy of J. N. Bahcall, from *Neutrino Astrophysics* (1989) Cambridge University Press; *30* Courtesy of the European Space Agency; *31* After D. O. Gough, courtesy of the European Space Agency (1985); *33* Painting by Davis Melzer; *35* © 1989, the Royal Observatory, Edinburgh; *36, 63* © Royal Observatory, Edinburgh; *37* © 1989, Mark McCaughrean and the Royal Observatory, Edinburgh; *41* After C. J. Lada, in *Ann. Rev. Astr. Astrophys.* (1985); *43* After B. T. Draine, in

Astrophys. J. (1983); *44* Courtesy of W. Dent and the Royal Observatory, Edinburgh (1990); *46* Courtesy of R. Perley, © NRAO (1985) and *Astrophys. J.* (1984); *47* (*a*) From *Structure and Evolution of Active Galactic Nuclei* (1986), D. Reidel and Company; (b) Courtesy of K. Pounds and the European Space Agency; *48* Courtesy of Dr S. J. Bell-Burnell; *50* After S. L. Shapiro and S. A. Teukolsky, from *Black Holes, White Dwarfs and Neutron Stars: The Physics of Compact Objects* (1983), Wiley-Interscience; *51* Courtesy of H. Tananbaum et al. from *Astrophys. J.* (1972); *56* After A. Wandel and R. F. Mushotsky from *Astrophys. J.* (1986); *57* After T. J. Pearson et al., from *Extragalactic Radio Sources* (1982), D. Reidel and Company; *59* After A. R. Sandage from *Observatory* (1968); *60* Courtesy Dr M. Seldner; 64 After C. Frenk, From *Phil. Trans. Roy. Soc. Lond.* (1986); *65, 66, 67, 72* From M. S. Longair, *Alice and the Space Telescope* (1989), Johns Hopkins University Press, drawings by Stephen Kraft; *69* From R. V. Wagoner, in *Astrophys. J.* (1973); *70* Adapted from J. Audouze, *Astrophysical Cosmology* (1982) Pontifica Academia. Scientiarium Scripta Varia, Vatican City; *71* After F. Close, M. Marten and C. Sutton from *The Particle Explosion* (1987), Oxford University Press; *73* From A. Guth and P. Steinhardt, *The New Physics* (1989), Cambridge University Press

Glossary

absolute zero The point at which all molecular motion ceases and so, theoretically, the lowest possible temperature. It is the zero point of the Kelvin temperature scale used in science.

absorption The process by which the intensity of radiation decreases as it passes through a material medium. The energy lost by the radiation is transferred to the medium.

active galactic nucleus (AGN) The central region of an active galaxy in which exceptionally large amounts of energy are being generated from a source other than the normal output of individual stars. This is a common characteristic of various types of objects that have been classified differently according to their appearance and the nature of the radiation they emit. Quasars, Seyfert galaxies and radio galaxies are all manifestations of the AGN phenomenon. The source of power appears to be concentrated within the nucleus. One mechanism that might account for the huge amount of energy observed is the presence of a supermassive black hole, into which matter is falling with the release of gravitational energy. Variability over relatively short time-scales shows that the power-house must be concentrated into a small region of space.

angular momentum The momentum a body possesses by virtue of its state of rotation. An object with angular momentum will continue to rotate at the same rate unless a force acts on it.

anthropic principle The assertion that the existence of intelligent life influences the nature of the Universe we observe. According to the anthropic principle, we are not passive observers of a Universe that would be the same whether we were here to observe it or not. Rather, the Universe is as we see it because we exist.

antimatter Matter composed of elementary particles that have masses and spins identical to the particles that make up ordinary matter, but with certain other properties, such as electric charge, reversed. Although some antiparticles are observed in nature and others are produced in the laboratory, there is no evidence for the existence of antimatter in bulk as such. If ordinary matter and antimatter were to meet, they would annihilate each other with the release of energy.

Big Bang A model for the history of the Universe that states it began in an infinitely compact state and has been expanding ever since a creation "event", which took place something between 13 and 20 billion years ago, and has come to be known as the "Big Bang". The theory is now widely accepted since it explains three of the most most significant observations in cosmology: the expanding Universe, the existence of the cosmic microwave background radiation and the origin of the light elements.

binary star A pair of stars in orbit around each other, held together by their mutual gravitational attraction. About half of all "stars" are in fact binary or multiple, though many are so close that the components cannot be seen individually. The presence of more than one star is inferred from the appearance of the combined spectrum. The two components in a binary system each move in an elliptical orbit around their common centre of mass. Some binary components are so close that the pull of gravity distorts the individual stars from their normal spherical shape. They may exchange material and be surrounded by a common envelope of gas. An accretion disc may develop as material streams towards a compact, spinning star in a binary system. The energy released results in the emission of X-rays. Novae are another consequence of mass exchange in binary stars.

bipolar flow Gas streaming outwards in two opposing directions from a newly formed star. The flow emerges from the centre of the star's accretion disc and is constrained to move along the rotation axis. The stellar wind, blowing out at typical speeds of 200 kilometres per second, sweeps up interstellar material before it, creating a double-lobed shell structure, extending outwards for a distance of about a light year.

black body A body that absorbs all the radiation incident on it. The intensity of the radiation emitted by a black body and the way it varies with wavelength depend only on the temperature of the body and can be predicted by quantum theory.

brown dwarf A name coined for hypothetical very cool stars with masses too low for nuclear reactions to be ignited in their cores. No such brown dwarf has yet been unequivocally identified though a number of candidates have been suggested.

collimation The process of making a beam of light or particles parallel so that it is neither converging nor diverging.

cosmic microwave background radiation Diffuse electromagnetic radiation that appears to pervade the whole of the Universe. Its discovery in 1963 was of immense importance in cosmology because it provided strong evidence in favour of the Big Bang theory. It is believed to be the relic of the radiation generated in the event that marked the origin of the Universe. The spectrum of the background radiation is characteristic of a black body at a temperature of 2.736 degrees above absolute zero (2.736 K) and is most intense in the microwave region. The Milky Way Galaxy is travelling through space at 600 kilometres per second relative to the background radiation.

critical density In cosmology, the density of matter that would ensure that the Universe just expands to infinity. The observed expansion can be slowed and even reversed by gravity if the density is sufficiently high. The critical density is defined as the density that would ensure the deceleration and the velocity of expansion becoming zero at the same time. Its value is $1-2 \times 10^{-26}$ kilograms per cubic metre. This is about ten times larger than the density inferred from visible matter, such as stars and galaxies. Many cosmologists like to imagine that the Universe is closed, which leads to the missing mass problem and the search for dark matter in order to match the actual density to the critical density.

dark matter Matter postulated to exist in the Universe but not yet detected. The evidence for dark matter stems partly from observations of the velocities of galaxies within galaxy clusters. Other evidence for dark matter comes from comparing the mass contributed by galaxies with that needed theoretically for a closed Universe.

deceleration parameter A number that indicates the rate at which the expansion of the Universe is slowing down due to the mutual gravitational attraction of the matter within it. A value greater than 0.5 denotes a situation in which the expansion of the Universe will ultimately be halted, followed by contraction and collapse. A value less than 0.5 indicates that the Universe will continue to expand. Present evidence tends to suggest that the expansion is slowing down.

degenerate star A term that covers both white dwarfs and neutron stars, which are made up of degenerate matter. These stars are in an advanced state of evolution and have suffered extreme gravitational collapse. Normal atoms cannot exist under the conditions of very high pressure. In white dwarfs, the electrons and atomic nuclei form a dense, compressed mass. A quantum-mechanical effect, called degeneracy pressure, counters further gravitational collapse. However, if the total mass of the star exceeds about 1.44 times the mass of the Sun, the degeneracy pressure is insufficient to balance the gravitational force and a neutron star results. Electrons and nuclei combine into a form of matter consisting of tightly packed neutrons.

density parameter In cosmological theory, the density of the Universe relative to the critical density.

dust grains Small particles of matter, typically of order 100 nanometres in diameter, which co-exist with atoms and molecules of gas in interstellar space. The dust grains are thought to consist mainly of silicates and/or carbon in the form of graphite. Dark clouds of dust are evident when they obscure the light from stars and luminous gas clouds, as happens in the plane of the Milky Way. Though tenuous, such clouds are very effective at absorbing visible light. Radiation at infrared and longer wavelengths can pass through the dust clouds unimpeded, however. The presence of dust is also revealed by the emission of infrared radiation, generated when the grains are warmed by the absorption of visible and ultraviolet radiation. The temperature of dust is typically in the range 30–500 K.

Eddington limit An upper limit on the ratio of the luminosity to mass of a stable star generating energy through the conversion of hydrogen to helium. Sir Arthur S. Eddington (1881–1944) showed that the limit is 40 000 when units of the Sun's mass and luminosity are used. If the limit is exceeded, the outer layers of the star are blown away by radiation pressure and a planetary nebula is formed.

escape velocity The minimum velocity that will enable a small body to escape from the gravitational attraction of a more massive one.

Euclidean space Space that is uniform, homogeneous and isotropic (i.e. all places and all directions are alike) and in which geometrical shapes are unaltered by rotation of position. Mathematically, the space has zero curvature. This is the space generally adopted for simpler cosmological models and is the "flat" space of everyday experience.

general theory of relativity A theory of gravitation, published by Albert Einstein in 1915, based on the geometry of four dimensions. This theory deals with accelerated motion, whereas the earlier (1905) special theory of relativity considered uniform motion. One of the fundamental postulates of the general theory is that it is impossible for an observer inside a black box to distinguish experimentally between uniformly accelerated motion and motion in a gravitational field. The general theory describes gravitation as a geometric property of four-dimensional space-time. This geometry is in turn determined by the mass and energy distribution. In the general theory, mass and energy are equivalent, being connected through the famous equation $E = mc^2$. The general theory of relativity features in several areas of astronomy. The perihelion of Mercury advances by 43 seconds of arc per century more than is predicted by Newton's gravitational theory but relativity explains it exactly. The light from stars is deviated if the ray passes very close to the Sun and this has been observed at solar eclipses. The motion of pulsars in binary systems is readily accounted for by the general theory of relativity. Its most frequent application is in cosmology. Gravity is an attractive force operating over great distances and so the theory of gravity (i.e. the general theory) is crucial to attempts to make mathematical models of the Universe.

giant molecular cloud A cloud of interstellar matter containing molecules. Giant molecular clouds are made up primarily of molecular hydrogen and carbon monoxide (CO) but they also contain many other interstellar molecules. They are the most massive entities within our Galaxy, containing up to 10 million solar masses, and are typically 150 to 250 light years across. The Orion Nebula is a notable example.

Hubble constant The constant of proportionality in Hubble's Law. It describes the rate at which the Universe expands with time. The value is not easy to determine, on account of uncertainties in the extragalactic distance scale, but it appears to lie between 50 and 100 kilometres per second per megaparsec. In an evolving Universe, its value changes with time, rather than being a true constant. For that reason, some prefer to refer to it as the Hubble "parameter".

Hubble's Law A statement to the effect that the recession velocities of distant galaxies are directly proportional to their distance from us. The law is an immediate consequence of the uniform expansion of the Universe.

hyperbolic space Space that is uniform in the same way as Euclidean space but with the additional character that there is more than one parallel to a straight line through a given point. Mathematically, the space has negative curvature.

inflationary Universe A class of Big Bang models of the Universe that include a phase change at about 10^{-34} seconds after the onset of the Big Bang. This event released enormous energy, stored until then in the vacuum of space-time. The horizon of the Universe expanded, temporarily, much faster than the speed of light. This theory is able to account satisfactorily for the present vast extent of the Universe and its uniformity.

ionization The process in which electrons are removed from atoms, through collisions for example or because the temperature is sufficiently high, leaving the atoms with a net positive electric charge. The electrically charged atoms become positive ions.

isotropy The property of having no preferred direction. Liquid water is isotropic whereas a snowflake, which has sixfold symmetry, is not. The Universe on the largest scales is believed to be isotropic.

M numbers, Messier catalogue A catalogue of about 100 of the brightest galaxies, star clusters and nebulae, compiled by the French astronomer Charles Messier (1730–1817). His initial list, published in 1774, contained 45 objects but it was supplemented later with additional discoveries and contributions from Messier's colleague, Pierre Méchain. Objects in the catalogue, which is still widely used, are identified by the prefix "M" and their catalogue number.

magnitude A measurement of the brightness of a star or other celestial object. On the magnitude scale, the lowest numbers refer to the objects of greatest brightness. The magnitude system was initially a qualitative attempt to classify the apparent brightness of stars. The Greek astronomer Hipparchus (c. 120 BC) ranked stars on a magnitude scale from "first" for the brightest stars to "sixth" for those just detectable in a dark sky by the unaided eye. This qualitative description was standardised in the mid 19th century. By this time it was understood that each arbitrary magnitude step corresponded roughly to a similar brightness ratio. (In other words, the magnitude scale is a logarithmic scale of brightness.) The brightness of stars or galaxies as observed from the Earth, i.e. their apparent magnitude, depends on both their intrinsic luminosity and their distance. Absolute magnitude is a measure of intrinsic luminosity on the magnitude scale. It is defined as the magnitude an object would have at the arbitrary distance of 10 parsecs.

main sequence In stars on the main sequence, the fusion of hydrogen into helium in the stellar core is the source of energy. The expression "main sequence" arises because if a sample of stars is plotted on a graph of stellar luminosity against stellar temperature, the hydrogen burning stars occupy a broad track with high luminosity/high temperature stars at one end, and low luminosity/low temperatures at the other. Cool stars with high luminosities, such as red giant stars, and hot stars with low luminosities such as white dwarfs are not on the main sequence.

neutrino astronomy The attempt to detect neutrinos from cosmic sources, especially the Sun. Neutrinos are elementary particles with no electric charge, which interact only very weakly with other matter.

They travel at the velocity of light and are produced in vast quantities by the nuclear reactions that take place in the centres of stars and in supernova explosions.

neutron star A star that has collapsed under gravity to such an extent that it consists almost entirely of neutrons. Neutron stars are only 10 kilometres across and have a density of about 10^{17} kilograms per cubic metre

NGC numbers, *New General Catalogue* A catalogue of non-stellar objects compiled by J. L. E. Dreyer of Armagh Observatory and published in 1888. It listed 7840 objects. A further 1529 were listed in a supplement that appeared seven years later, called the *Index Catalogue* (IC). *The Second Index Catalogue* of 1908 extended the supplementary list to 5386 objects. The NGC and IC numbers are widely used as a means of identification of nebulae and galaxies.

occultation The passage of one astronomical object directly in front of another so as to obscure it from view as seen by a particular observer.

planetary nebula An expanding shell of gas surrounding a star in a late stage of stellar evolution. The name derives from the description given by William Herschel, who thought their circular shapes to be reminiscent of the discs of the planets as seen through a small telescope. There is no connection between planets and planetary nebulae. Planetary nebulae are formed in the process of mass loss during which red giant stars ultimately become white dwarfs. The nebulae take a variety of forms - ring-shaped, circular, dumbbell-like or irregular. Notable examples include the Ring Nebula, the Helix Nebula and the Dumbbell Nebula.

polarisation The non-random distribution of electric field direction among the photons in a beam of electromagnetic radiation. In linear polarisation, the vectors representing electric field are parallel. In circular polarisation, the direction of polarisation changes continuously in such a way that the electric field vector rotates with the frequency of the radiation.

protostar A star in the earliest observable stage of the formation process when condensation is taking place in an interstellar cloud but before the onset of nuclear reactions in the interior.

pulsar A stellar source of radio waves, characterized by the rapid frequency and regularity of the bursts of radio waves emitted. The time between successive pulses is milliseconds for pulsars in binary systems and up to 4 seconds for the slowest. A pulsar is a rotating neutron star, with a mass similar to the Sun's but a diameter of only about 10 kilometres. The pulses occur because the neutron star is rotating very rapidly.

quasar A small extragalactic object that is exceedingly luminous for its angular size and has a high redshift. Quasars are now generally believed to be the most luminous type of active galactic nuclei (AGN). In a small number, the faint nebulous light of a surrounding galaxy has been detected. Many thousands of quasars are catalogued. In general, quasars have a spectrum that shows emission lines, high redshifts (typically from 0.5 to 4, although higher and lower values than these are recorded), and they are so compact that they appear as sharp as stars on photographs. Although the quasars discovered in the 1960s were all radio sources, the majority of those now known are not strong radio sources.

radio galaxy A galaxy that is an intense source of radio emission.About one galaxy in a million is a radio galaxy. The radio emission is synchrotron radiation from electrons travelling at speeds close to the velocity of light. The prototype is often regarded as Cygnus A, in which two huge clouds of radio emission, are disposed symmetrically on each side of a disturbed elliptical galaxy. Radio galaxies are closely related to quasars, many of which have similar radio properties.

Schwarzschild radius The critical radius at which the space-time surrounding a sphere becomes so curved that it wraps round to enclose the body. An object that has collapsed inside its Schwarzschild radius is a black hole, from which nothing can escape into the outside world. The Schwarzschild radius of the Sun is 3 kilometres and of the Earth, 1 centimetre.

Seyfert galaxy A type of galaxy with a brilliant point-like nucleus and inconspicuous spiral arms, first described by Carl Seyfert in 1943. The spectrum shows broad emission lines.

singularity A mathematical concept that can be visualised as a warped region of space-time where one or more components of the equations describing the geometry become infinite so that ordinary physical laws cease to apply. The Big Bang is thought to have emerged from such a singularity according to the classical picture.

special theory of relativity The theory postulated by Einstein in 1905. In it, Einstein assumed that the laws of physics are the same in all frames of reference that are in uniform motion relative to each other, and that the speed of light is a fundamental constant that has the same value for all observers. The theory has many important consequences: for example, the observed mass of an object increases with velocity and its observed length decreases.

superluminal motion Motion at a velocity that apparently exceeds that of light. The angular separation of the components of some double radio sources are increasing at a rate that is apparently equivalent to as much as ten times the speed of light when the distance of the source is taken into account. In reality, the effect is a purely geometrical one caused by the fact that the ejected component is travelling almost directly towards us along the line of sight at a velocity nearly as great as that of light. The phenomenon has been observed in the quasar 3C 273.

supernova A catastrophic stellar explosion in which so much energy is released that the supernova alone can outshine an entire galaxy of billions of stars. In addition to the light energy produced, ten times as much energy goes into the kinetic energy of the material blown out by the explosion and a hundred times as much is carried off by neutrinos.A supernova explosion occurs when an evolved star has exhausted its nuclear fuel. Under these circumstances, the core becomes unstable and collapses in less than a second. The rate accelerates as iron nuclei break up and neutrons form. However, implosion cannot continue indefinitely. When the density of nuclear matter is reached, there is strong resistance to further pressure, the core bounces and an outward shock wave is generated. The outer layers of the star are blown outwards at thousands of kilometres per second, leaving the neutron star core exposed.

supernova remnant The expanding shell of material created by the ejection of the outer layers of a star that explodes as a supernova.Well-known examples of supernova remnants are The Crab Nebula, Cassiopeia A, Kepler's Star, Tycho's Star and the Cygnus Loop.

synchrotron radiation Electromagnetic radiation emitted by an electrically charged particle travelling almost at the speed of light through a magnetic field. The name arises because it was first observed in synchrotron accelerators used by nuclear physicists. It is the major source of radio emission from supernova remnants and radio galaxies. Much of the light and the X-ray emission from the Crab Nebula is also produced via the synchrotron process by the very high energy electrons from the central pulsar. The spectrum of synchrotron radiation has a characteristic profile very different from that of the thermal radiation emitted by hot gas. Synchrotron sources are thus easy to identify. The polarisation of the emission provides a means of estimating the magnetic field in the source.

white dwarf A star in an advanced state of stellar evolution, composed of degenerate matter. A white dwarf is created when a star finally exhausts its possible sources of fuel for thermonuclear fusion. The star collapses under its own gravity, compressing the matter to a degenerate state in which electrons and nuclei are packed together.

WIMPs Acronym for electrically neutral weakly interacting massive particles. These have not yet been observed in particle accelerators. If they exist, they are important in cosmology because they would contribute to the missing mass.

Units

The following conventions are used for the units throughout this book. SI (Systême International) units are always used and so are based upon the metre as the unit of length, the kilogram as the unit of mass and the second as the unit of time. Some of the derived units used in the text are as follows.

Unit of wavelength In optical and infrared astronomy, it is conventional to use **nanometres (nm)**, meaning one billionth of a metre, i.e. 1/1 000 000 000 metre or 10^{-9} metre and **microns** (μ**m)**, meaning one millionth of a metre, i.e. 1/1 000 000 metre or 10^{-6} metre.

Unit of frequency The unit of frequency is one cycle per second which is known as **1 hertz (Hz)**. Radio frequencies are described in **Megahertz (MHz)**, meaning 1 000 000 hertz = 10^6 hertz, and **Gigahertz (GHz)** meaning 1 000 000 000 hertz or 10^9 hertz.

Unit of energy The unit of energy in the SI system is the joule but it is permissable to use the unit of the **electron-volt** to describe the energy of high energy **photons**, or the particles of electromagnetic radiation. For example, it is conventional to describe the energy of X-ray photons in terms of electron-volts (eV) or kiloelectron-volts (keV) which means 1000 electron-volts. Correspondingly, energetic gamma-rays are often referred to in terms of megaelectron-volts (MeV) meaning 1 000 000 eV.

Units of distance In astronomy, the conventional unit is the parallax-second which is abbreviated to **parsec (pc)**. It corresponds to a distance of about 3×10^{16} metre. A closely similar unit is the **light year** which is about one-third of a parsec.